Using Computers Effectively

Using Computers Effectively

How Teachers and Students Can Improve Their Knowledge and Practice

Debojit Acharjee

ROWMAN & LITTLEFIELD
Lanham • Boulder • New York • London

Published by Rowman & Littlefield
An imprint of The Rowman & Littlefield Publishing Group, Inc.
4501 Forbes Boulevard, Suite 200, Lanham, Maryland 20706
www.rowman.com

86-90 Paul Street, London EC2A 4NE

Copyright © 2023 by Debojit Acharjee

All rights reserved. No part of this book may be reproduced in any form or by any electronic or mechanical means, including information storage and retrieval systems, without written permission from the publisher, except by a reviewer who may quote passages in a review.

British Library Cataloguing in Publication Information Available

Library of Congress Cataloging-in-Publication Data

Names: Acharjee, Debojit, 1982- author.
Title: Using computers effectively : how teachers and students can improve their knowledge and practice / Debojit Acharjee.
Description: Lanham, Maryland : Rowman & Littlefield, [2023] | Includes bibliographical references and index.
Identifiers: LCCN 2023001323 (print) | LCCN 2023001324 (ebook) | ISBN 9781475868791 (cloth) | ISBN 9781475868807 (paperback) | ISBN 9781475868814 (ebook)
Subjects: LCSH: Microcomputers—Study and teaching. | Computer literacy.
Classification: LCC QA76.27 .A27 2023 (print) | LCC QA76.27 (ebook) | DDC 004.071—dc23/eng/20230222
LC record available at https://lccn.loc.gov/2023001323
LC ebook record available at https://lccn.loc.gov/2023001324

To my parents,
Dilip Acharjee
and
Makhani Bhuyan Acharjee,
for their relentless love, support, and encouragement.

Contents

Preface		ix
1	What's a Computer?	1
2	What Are Hardware and Software?	13
3	What's a CPU?	23
4	What's a GPU?	31
5	What's Computer Memory?	37
6	What Are Computer I/O Devices?	49
7	What's an Operating System (OS)?	59
8	What's Application Software?	65
9	What's a Computer Network?	73
10	What's the Internet?	83
11	How to Choose a Computer	93
12	How to Buy a Computer	101
13	How to Build a Computer	107
14	How to Install an OS and Apps on a Computer	121
15	How to Use a Computer for Homework and Teaching	131

Glossary	139
References	145
Index	147
About the Author	157

Preface

Computers have always fascinated me since my childhood and this fascination has made me learn about computers with great enthusiasm. Although I have learned so much about computers since my school days, I feel like there is no limit to this learning with the rapid change in technology. Computers have evolved a lot since I learned computers during my childhood. Earlier, when I was a kid, I used to learn BASIC and DOS on a computer with Intel i486DX2 processor and a CRT display. But nowadays I am amazed by the output quality and processing power of a desktop computer.

I believe that what we can do today with computers was not possible a few decades ago. As the price versus computing power of desktop computers became reasonable, using a desktop computer for everyday use has become possible for everyone. When I was in college, there wasn't much internet speed and computing power to do much like HD gaming and video streaming, which have become as easy as pie these days.

That's why I thought of using all my knowledge about computer hardware and software to write this book. I have written this book in such a way that it will give overall knowledge about computers and teach about buying and building a computer. This book contains all the fundamentals related to computer hardware and software that could be useful for almost anyone who is interested in learning computers. However, the last chapter of this book is mainly for students and teachers. This chapter gives information about using various types of apps and internet resources for doing homework and teaching.

Being a computer professional and an enthusiast, I have used computers for all kinds of work. I learned many ways to use computers for various purposes and realized that a perfect combination of software and hardware makes a computer suitable for a specific job. I have included all these nuts and bolts about computers in this book, and I hope students and teachers will be able to improve their knowledge from this book.

In this book there are many abbreviations used in most of the chapters. To know about their meaning and full form, please refer to the glossary.

This book is not intended for professional assembling and manufacturing of desktop computers, but for learning how to build a computer for educational purposes only.

DEBOJIT ACHARJEE

Chapter 1

What's a Computer?

A computer is a machine that can process data using arithmetic or logical operations based on instructions known as programs. A computer is a machine, and like any other machine—electronic or mechanical—it does some work. For example, a washing machine is a machine that washes clothes, a dish washer is a machine that cleans dishware, and a bike is also a machine that lets people travel, but it's a mechanical one. So what does a computer do being a machine?

A simple answer would be that it processes data. Now what exactly does data processing mean? To better understand that—first understand what data are exactly. Any information can be considered as data, but data can be of various types like text, image, audio, video, and so on. Even text data can be numeric and character. So a computer is a machine that works on these types of data to either store them or make any changes (using arithmetic operations), which is called data processing.

COMPUTER IS A LOGICAL MACHINE

But a computer can do any such data processing using logical and arithmetic operations only, which is similar to a digital calculator—used for only mathematical operations. A simple calculator is also a simple computer, but it processes only numeric data and it can do basic arithmetic operations like addition, subtraction, multiplication, and division. But whether it's a computer or a calculator, none of these digital machines can understand any data types except binary—either 0 or 1.

But what exactly is binary data? This is not exactly conventional data but limited to the computer only. This is something like "1" for YES and "0" for NO. So the question is, why and where are this YES/NO data used by a computer?

A modern computer is a digital machine, which means it uses digital electronics, and in any digital electronic device the binary data is used for various operations. Any digital electronic device consists of a circuit (the internal wiring of any electronic device). Just like any house with electrical appliances has electrical wiring—connecting the lights, fans/AC, heaters, and so on, to the switches. Whenever any appliance needs to be turned on, the relevant switch needs to be switched on, and can be turned off by switching it off.

Similarly, in any digital circuit such switching of components is done automatically and when it's ON it means "1" and OFF means "0." Inside a computer also such a kind of switching is done, and just like how electricity flows through the wire of any connected light bulb, when it's turned on, electricity also flows through the wires on the circuit board of a computer, which are known as buses. When any electricity flows through the bus—it's taken as "1" by the computer and when little or no electricity flows—it's taken as "0".

In this way 0's and 1's flow through the circuit of a computer representing binary data. But this binary data is actually converted from the source data, which could be text, image, audio, video, and so on. As human readable data is different from the binary data—any type of data has to be converted to binary type, so that it can be used by the computer for processing.

Now, what kind of processing is done by a computer? A computer can only understand binary data and can do mathematical operations with such binary data only, which is also known as Boolean operations.

Based on such Boolean operations, a computer can process the data and carry out various operations like show the processed data—just like a calculator shows the calculated result (by converting the binary data to human readable data type) or can store the data on a storage device like a hard disk.

COMPUTER NEEDS INSTRUCTIONS

However, a computer cannot work like any ordinary machine and needs instructions to carry out any operation. Such instructions come from a computer program—code of instructions that follow logical operations. A computer program is programmed by a programmer using a computer language like BASIC, C/C++, Python, and so on. These are also known as high-level languages and the binary data used by the computer is called low-level/machine language. Just like a driver drives a car but has no idea about what's happening inside the engine and the gearbox, but steers the wheel and changes the gear to drive the car.

Similarly, a computer program runs the computer without knowing anything about what's happening inside the circuit. A program gives instructions to the computer about what to do and what not to do. In computer science the circuit of a computer is called hardware and the computer program is called software. The relation between hardware and software makes a complete computer system.

COMPUTER WORKS WITH HARDWARE COMPONENTS

A computer hardware is any physical part of a computer that is responsible for its operation. For example, the circuit board of a computer—consisting of various electronic components—is a computer hardware. Apart from the computer itself, there are two types of computer hardware that are also required for the computer to make it complete—an input and output device. Input devices like the keyboard and a mouse are used to enter data into the computer and without an input device it's not possible to communicate and feed data to a computer.

Even for playing a video game an input device is used and a joystick is one such standard input device used for playing video games on a computer and also in many video game consoles like Nintendo and PlayStation.

However, many video games can be played on a computer using only a keyboard and mouse. Apart from that, a microphone and a digital camera are also input devices that fall under computer hardware. With the help of a microphone, it's not only possible to record sound digitally but can also be used to give voice commands to a computer.

A digital camera can be connected to a computer to record video or take pictures that can be stored digitally, and it's also possible to make live video broadcast using a digital camera and can be viewed on another computer connected to the broadcasting computer though a communication medium like local network or internet.

Besides the input and output devices, a computer may use a communication device like a modem to connect with other computers, and any such communication device also falls under computer hardware.

Besides a modem, a scanner is another input device like a digital camera that is used for converting a hard copy of any document to a soft one that can be stored digitally using a computer. It's also one kind of computer hardware.

Just like the input devices, there are many output devices that are used to connect with a computer to get output data (data after being received and processed from an input device). One such important output device is the computer monitor or display unit. Without a computer monitor, it's not easy to operate a computer. Just like without the LCD display of a digital calculator or a watch it's not possible to see the calculated result or the time without a display—it's also not possible to operate and use the computer.

Earlier, the computer monitors were made from cathode-ray tubes (CRT), but now they are LCD/LED displays. Any such display device used as an output device is a computer hardware.

Speakers used with a computer are also output devices and are used to play sound from a computer. Any computer-generated sound is digital and uses a sound card to reproduce the sound through the speakers. Computer speaker systems come in 2.0, 2.1, 4.1, 5.1, 7.1, and so on. This format represents the number of speakers and the subwoofer. The first number represents the number of speakers and the second number represents the subwoofer.

So a 2.0 speaker system means two speakers with no subwoofer and 2.1 means with one subwoofer. A subwoofer is used to reproduce low frequencies (bass). Sound systems starting from 4.1 are used for surround sound and it means four speakers—two front and two back with one subwoofer. Similarly 5.1 and 7.1 sound systems use more speakers to create more realistic surround sound effects. Any such sound system used to reproduce computer-generated sound is a computer hardware.

Printers are also output devices used to print documents, and they are mostly used in offices to print documents and reports created by a computer. Printers can print either black and white or color documents on plain paper. Printers can be dot-matrix, inkjet, or laser based, and the printing quality varies depending on the type of printer. A dot-matrix printer can print only monochrome documents with text only and is a slow one. Inkjet printers use ink to print documents and pictures.

Color inkjet printers use Cyan, Magenta, and Yellow inks to print color photos along with the black ink, and uses only black ink for black and white printing. Laser printers on the other hand use colored powders called toners to print tests or images on papers using laser beams. Any such printers used with a computer falls under computer hardware.

Nowadays the modern computer display that comes with a touchscreen can be considered as an input device and also an output device because a touchscreen can take inputs from the user's touches on the screen.

A computer uses a storage device like a hard disk to store data permanently because a computer memory inside the circuit board requires power to hold the data, and it gets wiped out after the computer is powered off. Any such storage device falls under computer hardware, which includes many other storage devices like flash drives, CD/DVD drives, and the very old tape drives. Any storage device works like an input device and also as an output device because data is read and written to a storage device by the computer, and it's a part of a computer hardware.

COMPUTERS HAVE AN ARCHITECTURE

Just like a house or a building consisting of many things like doors, windows, stairs, roof, and so on, make its architecture—a computer also has an architecture that consists of many things like input/output devices, I/O interface, Central Processing Unit (CPU)—including Arithmetic Logic Unit (ALU) and Control Unit (CU), Graphics Processing Unit (GPU), and memory and storage device(s). All these components are connected with each other through electrical wires that are embedded on the circuit board of a computer, which are called bus.

Binary data flows though the bus, and transfer of data from one component to another is done through a bus at a measurable rate, which is called transfer rate—measured in bits or bytes per second. Computer buses also have speed that is measured in hertz. Higher the transfer rate and speed of a bus, better is the speed and performance of a computer. Besides the bus speed, the CPU and memory are also responsible for the computing performance.

Just like the transfer rate, CPU has a speed and it's measured in hertz; higher the hertz of the CPU, better is its computing power. Computer memory is measured in bytes and higher the bytes of memory, better and powerful is the computer. Even though the GPU is only for graphics processing, its speed is also measured in hertz and it also requires memory. So higher the speed and memory of the GPU, better is its ability to process graphics (see figure 1.1).

The input device sends data to the CPU where it's processed with the help of the ALU, and data is temporarily stored in the memory. The processed data is then sent to the output device. When the data needs to be stored permanently, it's sent to a storage device. The GPU processes any graphics data and sends it to the output device like a computer display. The CU controls all the units including the I/O interface that helps transfer data between the CPU/GPU and the input/output devices.

There are two types of computer memory—Read-Only Memory (ROM), used to read data that is permanently stored, and Random-Access Memory (RAM), used to read/write data for temporary storage. The data stored in the RAM gets wiped out when the power is switched off. That's why RAM is also called volatile memory and other storage devices like ROM or hard disk that can store data permanently without power are called non-volatile memory. RAM is also called primary memory and other storage devices like hard disks are called secondary memory.

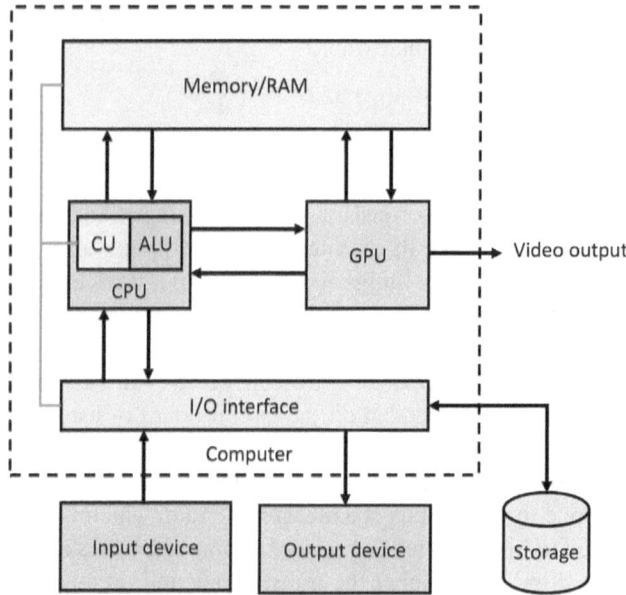

Figure 1.1 Basic Architecture of a Computer. Author Created.

TYPES OF COMPUTERS

A computer can be mainly of two types depending on its purpose of use. Computers designed for special purposes are called special-purpose computers. Such a computer can do only one kind of task and cannot be customized to do multiple types of tasks in various environments. For example, a calculator can be considered as a special-purpose computer and it can only do addition, subtraction, multiplication, and division. A calculator can't be used to play video games or play music like a piano.

However, some high-end graphics calculators can be programmed to play games, which a basic calculator can't. Besides that there are many home appliances like microwave ovens, washing machines, refrigerators, air conditioners, and so on, which come with a special-purpose computer. That's why the computer inside these machines can't be used for any other purpose.

On the other hand, a computer that can be used for multiple uses is called a general-purpose computer.

The best example of a general-purpose computer would be a desktop computer—also known as a home/personal computer. With a desktop computer it's possible to do multiple tasks like sending email, playing video games, watching videos, making 2D/3D drawing, playing music, and so on. With the help of software it's possible to make a computer work like a specific machine—a video player software can make a computer work as a media player, and a game emulation software can make a computer play Nintendo and PlayStation games. So a general-purpose computer can be converted to any electronic device with the help of a software.

However, depending on the hardware configuration and computing power, a general-purpose computer has many uses and can be of many types:

- Microcomputers—use microprocessor chips as a CPU and that's why the word "micro" is used to represent these types of computers. They became popular in the twentieth century, and Altair 8800 was one such popular microcomputer in 1975. Desktops, Laptops, Tablets, and so on, are microcomputers, and a desktop computer is the oldest and most popular type of microcomputer used worldwide.
- Minicomputers—bigger in size but more powerful than the microcomputers, which initially used transistors instead of microprocessors in the 1960s. The PDP-8 was the first minicomputer introduced by Digital Equipment Corporation's (DEC) in 1964. Such computers were used for scientific and engineering computations, business transaction processing, file handling, and database management.
- Mainframe computers—designed to be powerful enough to process large quantum of data from multiple users, connected to a single-user machine like a desktop. The term *mainframe* was derived from the large cabinet, called a *main frame*, which housed the central processing unit and main memory of early computers.

  As these types of computers can handle and process large amounts of data quickly, they are often used in large organizations for bulk data processing, such as census, industry and consumer statistics, enterprise resource planning, and transaction processing.

  Their processing power is measured in MIPS (million instructions per second) and can respond to millions of users at a time. System/360, z9, z13, and z15 of International Business Machines (IBM) are some examples of mainframe computers.
- Supercomputers—work faster than any other types of computer and use multiple CPUs to process data using a technique called parallel processing. Super computers are used for weather forecasting, fluid dynamics, nuclear simulations, theoretical astrophysics, and complex scientific computations. The speed of a supercomputer is measured in floating-point operations per second (FLOPS)—higher the value of FLOPS, more powerful is the supercomputer.

In 1960, the UNIVAC LARC was a mainframe computer built to do multiprocessing, and it was used by Edward Teller (inventor of hydrogen bomb) to run hydrodynamic simulations for nuclear weapon design. Today it is considered as the first supercomputer of that time and it was capable of doing 250 kilo instructions per second (kIPS). In 2022, HPE Cray built a supercomputer called Frontier (OLF-5), and it's capable of doing 1.102 EFLOPS (ExaFLOPS), which is considered as the fastest supercomputer of 2022.

HISTORY OF COMPUTERS

Before the emergence of electronics, computing devices were mechanical and didn't require any electricity. Abacus was one such computing device, and it's one of the oldest calculating devices used during the ancient times—Sumerian abacus was the first abacus used between 2700 and 2300 BC.

However, the Difference engine is considered as the first mechanical computer created by Charles Babbage in 1820, and it was designed to tabulate polynomial functions. Later he designed the Analytical engine in 1837, which was capable of taking program and data inputs through punch cards. It would output the data on a printer or plotter and could store data on punch cards. This mechanical computer had an ALU, control flow, and memory like a modern computer, and it's considered as the first general-purpose mechanical computer.

Even though the Analytical engine worked like a modern computer, it still was not programmable like a modern computer, and lacked the versatility and accuracy of modern digital computers. However, before the arrival of digital computers, another form of a computer called analog computers showed some improvements over mechanical computers like the Analytical engine. One such analog computer was the Tide-predicting machine invented by Sir William Thomson in 1972. This mechanical computer used wheels and discs to work as an analog computer.

It was capable of predicting the ebb and flow of sea tides, and could be programmed like a modem computer.

After the invention of electricity and electrical components like relays and switches, the design of electromechanical analog computers became possible. The US Navy had developed an electromechanical analog computer called the Torpedo Data Computer in 1938, which was small enough to use aboard a submarine. This computer used trigonometry to calculate and determine the firing accuracy of a torpedo at a moving target.

The concept of digital computers became popular after the creation of the Z2 computer by a German engineer Konrad Zuse in 1939, which was one of the first electromechanical digital computers that used electrical switches and relays. This was followed by the Z3 in 1941, and it was the world's first electromechanical automatic digital computer.

After the invention of vacuum tubes, the digital computers became free from electromechanical components, as the vacuum tubes were used in the computers instead, and this was the beginning of the modern electronic digital computer. In 1942, John Vincent Atanasoff and Clifford E. Berry of Iowa State University developed the first automatic electronic digital computer called Atanasoff–Berry Computer (ABC), which used about 300 vacuum tubes, with capacitors fixed in a mechanically rotating drum for memory.

Another digital computer called Colossus was developed by the British codebreakers in the year 1943, and it was the world's first electronic digital programmable computer. Later, another digital computer called Electronic Numerical Integrator and Computer (ENIAC) was built in 1945, and it was the first electronic programmable computer built in the United States. Both ENIAC and Colossus were similar, but ENIAC was much faster and flexible.

After the invention of transistors (made from semiconductor), vacuum tubes were replaced by transistors in 1955. Due to the smaller size of transistors, the size of modern digital computers reduced dramatically. On June 21, 1948, the digital computer called Manchester Baby, built at the University of Manchester in England by Frederic C. Williams, Tom Kilburn, and Geoff Tootill, was the world's first transistor-based computer to digitally store data and program on a storage device. Since then many other computers were built that used transistors, including Harwell CADET of 1955.

After the advent of Integrated Circuit (IC) and microprocessor, it became possible to build smaller and portable computers with more computing power. The first working ICs were invented by Jack Kilby at Texas Instruments and Robert Noyce at Fairchild Semiconductor. With the help of ICs and microprocessors various types of computers were built since 1960, including the mainframes (IBM 7090, IBM 7080, IBM System/360, BUNCH), minicomputers (HP 2116A, IBM System/32, IBM System/36, LINC, PDP-8, PDP-11), and microcomputers (Xerox Alto and Star, Altair 8800, Commodore 64, Apple I).

After the advent of the system-on-a-chip (SoC) ICs many companies were able to integrate the whole computer on a single chip and in 1992, Acorn Computers produced the A3010, A3020, and A4000 range of personal computers with the ARM250 SoC. Nowadays it has become possible to manufacture portable computers like tablet PCs and laptops with the help of SoC only.

DESKTOPS AND WORKSTATIONS

Among the microcomputers, desktops and workstations are most commonly used for home and business. However, there is a difference between these two computers—a desktop computer is similar to a workstation, but it's less expensive than a workstation and also has less powerful hardware. That's why workstations are mainly used for productive and research purposes. Most workstations come with multiple CPUs and GPUs to process high-quality graphics for video production or to run various simulation software.

On the other hand, desktop computers are used for simple productive work like video editing and also for gaming. But a desktop computer can be used at home for simple tasks like web browsing, emailing, social networking, and so on, as well as for business work like 3D drawing, video/audio production, software development, and so forth. That's why a desktop computer can be customized as per the needs, and the price of a desktop computer depends on its system requirements.

A desktop computer can be either purchased from a vendor as per the system requirement (CPU speed, memory, graphics type, etc.), and the price is determined by the hardware configuration (system requirements) and the brand. This method is also applicable for workstations too, but a workstation is always priced higher than a desktop because of the high-end hardware configuration. That's why it's possible to get a desktop or a workstation at a reasonable price when it's custom built (see figure 1.2).

Earlier, desktop computers used to come in a case that lay flat on the desk, but nowadays most desktop computers come in a case that is tower type and stands upright. The case of a desktop computer or a workstation encloses the circuit board (also called a motherboard—holds the CPU, RAM, and GPU all together), the power supply unit, storage devices like hard disk, and other peripheral devices. On the desk, a monitor is connected and used as an output device.

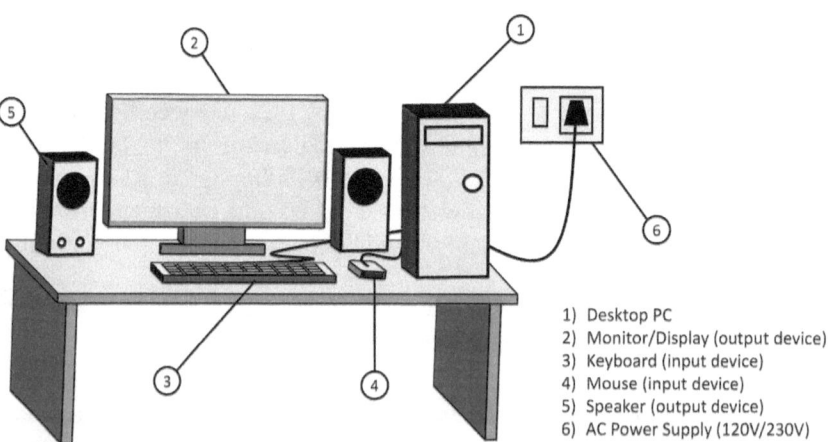

1) Desktop PC
2) Monitor/Display (output device)
3) Keyboard (input device)
4) Mouse (input device)
5) Speaker (output device)
6) AC Power Supply (120V/230V)

Figure 1.2 Input/Output Devices of a Desktop Computer. Author Created.

Earlier, CRT monitors were used, but nowadays LCD and LED monitors are most commonly used. Keyboard and a mouse are used to connect with a desktop computer as an input device, and with those, it's not easy to operate a desktop computer. A pair of speakers are at least required to get stereo audio output from the sound card (a peripheral device of a computer that generates sound) of a desktop computer.

Nowadays the sound card comes with the motherboard of a desktop computer, but a separate high-quality sound card can also be installed separately on the motherboard. Similarly, a separate high-end graphics card can also be installed on the motherboard to get better video quality, and that is usually helpful for playing games or video editing.

To power a desktop computer or workstation, AC power of 120V/230V (120V for US) is required and needs to be connected to the power outlet of an office or home. Even though desktop computers are less expensive than the workstations, both are good at consuming power and consume more than a 100W light bulb (except the newer types of low-powered smaller desktops and laptops). Just like an incandescent light bulb, a computer also heats up while working.

That's why the cases of desktop computers and workstations come with fans that keep everything cool inside. A better cooling system of a computer not only keeps the system cool but also improves the overall performance. Besides that, a desktop computer can malfunction or get damaged from static electricity and conductive liquid. So the case should be made from conductive material like metal (to protect it from electrostatic discharge) and should entirely cover the motherboard and the other connected peripheral devices, so that everything is protected from conductive liquid spills like impure water.

Therefore, a desktop computer or a workstation should be used as per the requirements and can be either purchased from a store or can be custom built using hardware parts, which also becomes budget friendly.

Chapter 2

What Are Hardware and Software?

Both hardware and software make a computer work properly. Either of these two things can make a computer perform good or bad. Hardware is basically considered the lowest layer of a computer, and it's the physical layer that comprises all the electronic components like the CPU, GPU/Graphics card, Memory/RAM, Storage devices, and the list goes on.

The software is the upper level that stays above the physical layer (hardware) of a computer. In a general-purpose computer, the software layer comprises an Operating System (OS) and applications. The OS consists of a kernel and the user interface. It's the software that gives instructions to the hardware and makes a computer work.

HARDWARE

All the hardware components of a computer are connected with each other using some kind of wiring called a bus. The bus is a bunch of wires or lines of conductive material (usually copper) printed on a circuit board (motherboard) that connects the CPU with all the other components like memory, GPU/ graphics card, and I/O interface. Without all these hardware components a computer can't work properly and the connection of all the hardware components on the motherboard makes a complete functional computer system.

- CPU—the only component that is directly or indirectly connected to every hardware component of a computer. That's why it's also known as the heart of a computer. Without a CPU, a computer can't work, and it's only the CPU that makes a computer complete. In a CPU there lies an ALU and the CU, but some modern CPUs also come with an integrated GPU.

The ALU does all the arithmetic operations like addition and subtraction, but only in binary format. The CU controls all the other components of the CPU and the computer. The FPU is used in modern CPUs to carry on floating-point operations and can perform various transcendental functions such as exponential or trigonometric calculations.

- GPU—used for graphics processing and most modern CPUs have an integrated GPU that helps render graphics without the need of any additional graphics processor. It's the GPU that makes it possible to play video games on a computer, and any application that has a graphical user interface (GUI) needs a GPU to run properly.

  A separate GPU can be used to improve the performance of the computer and that's why a discrete graphics card is often used in a video gaming computer. A discrete graphics card works independently, and that makes the CPU free from graphics processing load, which lets the CPU free to do other processing faster.

- Computer memory—part of a computer responsible for storing data used by the CPU. A computer memory can be either temporary or permanent. A temporary memory is used by the CPU to store data temporarily when the computer is powered on, and the data is wiped out when the computer is powered off. That's why this type of memory is also called a volatile memory. An example of such commonly used volatile memory is the RAM.

The RAM chip can store data only with the help of electricity, and as long as the chip receives electrical power, the data will remain stored, but as soon as the power is disconnected from the chip, the data gets wiped out. On the other hand, a permanent memory can hold data even when the computer is turned off. That's why this type of memory is called a non-volatile memory. ROM chips, hard disk, magnetic tapes, optical disc, and so on, are examples of non-volatile memory. A ROM chip can store data permanently and doesn't require any electricity to retain the data.

A hard disk can store data using an electromagnetic recording process on a magnetic (metal-coated) disc and this process is somewhat similar to the audio recording on a cassette tape, but in case of a hard disk the data is stored digitally (binary format). Instead of analog—unlike a tape recorder. As magnetic tapes and discs don't require electricity to hold the data, anything can be recorded on them permanently.

However, optical discs also record data in a similar process but it uses a laser instead of a magnetic field. The laser burns the surface of a disc and records the data digitally. The digital data is recorded on an optical disc in the form of 1's and 0's and the depth of a burn represents either 1 or 0. As the optical disc doesn't require any electricity to hold the data, anything recorded on it can become permanent. An optical disc can hold data as long as 100 years (theoretically).

- I/O interface—lets a computer connect with I/O devices. A computer requires input/output (I/O) devices to give instructions and to get the processed data. A computer uses an I/O interface to connect with all the I/O devices. Input devices like keyboard, mouse, joystick, and so on, need to be connected to a computer to operate the computer, and output devices like the monitor/display and a printer need to be connected to a computer to get the processed data.

  But it would become difficult for the CPU to directly handle too many I/O devices at the same time. That's why most modern computers use a special chip called the I/O chip or the I/O interface that communicates with the connected I/O devices. The I/O interface transfers data between the CPU and the I/O devices. Most modern computer motherboards use a super I/O chip to work as an I/O interface.
- Motherboard—the main circuit board of a computer that contains all the electronic components including the memory and CPU. The CPU is connected with the motherboard using a bus, which was earlier called as Front-Side Bus (FSB), but now it's called the HyperTransport/Uplink in AMD chipset motherboard, and Intel's Direct Media Interface (DMI) 2.0 or QuickPath Interconnect (QPI) in Intel chipset motherboards.

  Higher the speed of the CPU bus, better is the computer performance. Even the memory of a computer on the motherboard is connected with the CPU using a bus, and a higher memory bus speed could make a CPU perform better. On the other hand, a slower bus speed or a busy bus could cause a computer to slow down and when that happens, it's called a bottleneck.

That's why a motherboard designed for high-end CPUs will have a better bus speed and only higher bus speed can support CPU and RAM with higher clock speeds. Earlier, a motherboard used two chips—north bridge and south bridge—to connect all the components of the computer. The north bridge chip connects the CPU with RAM and graphics card, while the south bridge connects the I/O devices.

But now this design has changed, and nowadays most motherboards would come with only one chip (like a south bridge) to connect the I/O devices, and the CPU is connected with the RAM and graphics card through a dedicated bus (see figure 2.1).

The motherboard comes in various sizes, which is called the form factor. A motherboard form factor could be ATX, Mini-ATX, Micro-ATX, Mini-ITX, and so on, where ATX is the biggest size. Motherboards come with a chipset, and every chipset has a number. Usually, there are two types of chipset that makes the motherboards fall into two categories.

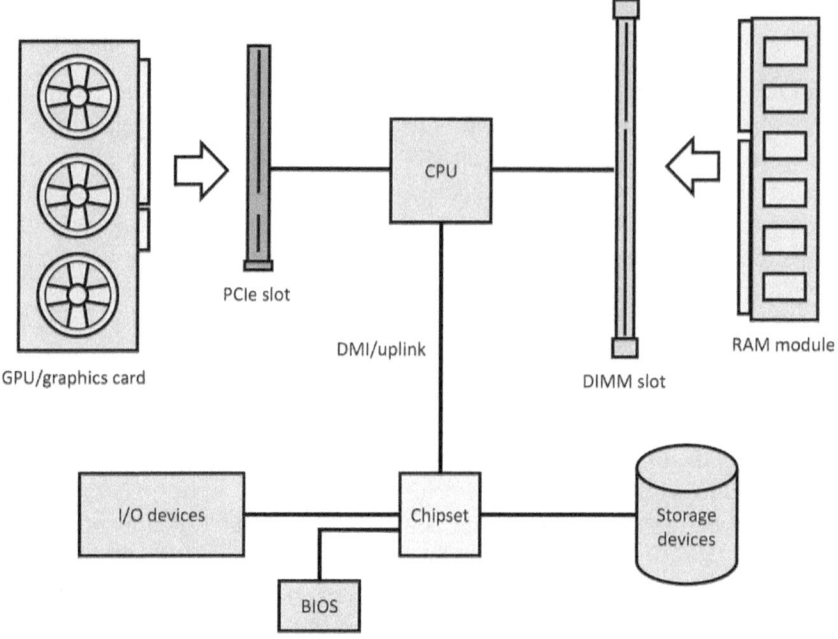

Figure 2.1 Block Diagram of a Modern Motherboard. Author Created.

There are main CPU manufacturers but only two types of CPUs are most commonly used in general-purpose computers, and the two general-purpose CPU manufacturers are Intel and AMD. That's why the chipsets for the motherboards are also made for either AMD or Intel CPUs, and most motherboard manufacturers need to design motherboards with a chipset either for AMD or Intel.

A motherboard with an AMD chipset can only support an AMD CPU and the one with an Intel chipset can only support an Intel CPU. A motherboard also has a dedicated socket for either AMD or Intel CPU, and either of them differs from one another and are not compatible with each other, which means that an AMD CPU can't be installed on a motherboard that supports Intel CPU and vice versa.

A motherboard also contains a chip called the Complementary Metal-Oxide-Semiconductor (CMOS) or the Basic Input/Output System (BIOS) chip. This chip holds the software (instructions) on how to start up the computer and helps the OS to communicate with the computer. Most modern motherboards require 3.3V, 5V, and 12V to operate and a separate power supply unit needs to be connected to the motherboard that provides these three types of voltages, and needs an AC 120V/230V power source (electrical outlet) to work.

A motherboard is enclosed in a metal case that is earthed (grounded) to avoid any damage to the electronic components from electrostatic discharge (ESD). Bigger motherboards need extra cooling fans inside the computer case to keep the temperature from rising. Although a CPU comes with a cooling fan, the computer case also requires extra fans to keep the temperature of the components of a motherboard normal. A better cooling system can keep the CPU running properly but overheating could affect its performance.

- I/O devices—used to operate a computer and get the output data processed by the CPU/GPU. Devices like keyboard, mouse, and joystick are used to operate a computer and input data to the computer, and that's why such devices are called input devices. A keyboard is used to enter any type of data (containing numbers and characters) to the computer or operate it using commands or keys.

A keyboard contains keys that work like a switch, and when a key is pressed, the chip on the circuit board of the keyboard sends a digital signal to the computer in the form of binary data, which is called scancode. The computer then maps this scancode to the key value sent by the keyboard. There are many types of keyboards but a standard keyboard used with most general-purpose computers is called the QWERTY keyboard, and a typical US keyboard contains 104 keys, but some other types of keyboards contain 105 keys.

On the other hand, a mouse is used to operate a computer with a GUI OS or application. A mouse sends digital data to the computer about its position representing a 2D plane of X and Y values. It also sends the mouse clicks and wheel movement signals to the computer. A mouse contains a left button, right button, and a wheel/middle button.

Some mice also contain additional buttons that are used for gaming or productivity. A joystick is an input device used for gaming and it sends signals to the computer about the stick position (like a mouse) and the buttons pressed while playing a video game. All these input devices can be connected to the computer using either a wired connection with the Universal Serial Bus (USB) port or a wireless connection using Bluetooth or Wi-Fi. However, some older keyboards or mice use a PS/2 port for connectivity.

A microphone can also be used as an input device to control a computer using voice recognition but it requires a high-end computer system preferably with Artificial Intelligence (AI). A document scanner and a webcam (camera) are also input devices, but can't be used to operate or control a computer. However, a webcam can be used to authenticate a user before operating a computer using face recognition technology.

A computer monitor is an output device used to view the output data of a computer. Earlier, computer monitors were called CRT displays because they used a CRT technology like a traditional television set. But nowadays Liquid Crystal Display (LCD) or Light-Emitting Diode (LED) displays are used instead of CRTs.

LCD uses liquid crystal technology to display texts or images but it requires the presence of light. That's why most LCD displays use fluorescent lamps on the sides of the LCD to make things visible on the LCD screen. On the other hand, LED displays also use liquid crystal technology, but it uses LEDs instead of fluorescent lamps, and consumes less power.

Besides these two types of displays, in-plane switching (IPS) and active-matrix organic light-emitting diode (OMLED) displays are also used, and have better picture quality than the LCD/LED displays. Earlier, most CRT displays had an aspect ratio of 4:3 but now most modern computer displays come with 16:9 aspect ratio with Full High Definition (FHD) image quality. Older displays could be connected to a computer using the Video Graphics Array (VGA) connector but new displays require a High-Definition Multimedia Interface (HDMI) port.

A printer is also an output device but it can't be used to operate a computer because a printer can't show the user interface activities in real time. That's why a printer is only used to print any data that is processed by the computer, and documents are usually the most commonly printed using a printer. A printer can be either dot-matrix, inkjet, or a laser and can be connected to a computer using an I/O port like parallel or USB port. However, most modern printers can be connected to a computer wirelessly using a Wi-Fi router.

Computer speakers are also used as an output device but they are mostly used for multimedia applications. A speaker audio system can be of 2.0, 2.1, 4.1, 5.1, and 7.1, where the number before the dot indicates the number of speakers and the number after the dot represents the number of subwoofers used for low frequencies (bass). A basic pair of computer speakers don't come with a subwoofer and that's why it's a 2.0 audio system.

More than two speakers are used to create surround sound effects, and that's why audio systems starting from 4.1 are used for gaming to experience surround sound audio. Besides that most motherboards come with an inbuilt speaker that gives audio beeps during startup, and the beep codes are used for troubleshooting any problem when the computer doesn't start up or power on.

Although multimedia speakers are not used as an output device to operate a computer, it can be used in some ways to interact with a computer. Computer speakers can be used as an accessibility tool for disabled people. Some OS comes with a feature called narrator that can read out text shown by the computer as an output.

Moreover, some computer systems with AI can talk to the user with the help of speakers, and independent talking speakers like smart speakers, such as Amazon Echo and Google Nest, can talk to the user. Therefore, it's possible to use speakers to operate a computer instead of a display unit.

Non-volatile storage devices like hard disks and memory cards that are connected through the I/O interface can work both as an input device and also as an output device. When data is sent to a storage device for storing, and during the write operation, it works like an output device, and while reading data from a storage device, it behaves like an input device—sending input data to the computer.

That's why when any stored computer program is executed from a storage device, it can give instructions to a computer instead of a user, and can operate a computer automatically, which is also called automation. Hard disk (mechanical and solid state) and CD/DVD drives require a Serial ATA (SATA) connection with the newer motherboards and older ones require Parallel ATA (PATA).

However, memory cards can be either used with a card reader (connected to the USB port) or a card slot permanently attached in some motherboards (especially in laptop motherboards). Usually a solid-state drive (SSD) and flash drives like the memory card are much faster than a mechanical drive, and can transfer data at high speeds.

That's why most new motherboards come with M.2. (NVMe) slot(s) for using these types of faster SSDs. This is mostly required for a gaming computer, and using such faster SSDs can improve the performance of a computer.

The overall performance of a computer mainly depends on the CPU and memory, but other components like the GPU and the I/O devices can also be responsible for the computer to work as expected.

SOFTWARE

A software is a set of programs that gives instructions to a computer. Without software a computer is just a piece of unusable machine. It's the software that makes a computer smarter and reliable. Any software is developed using a development kit with a programming language like C/C++ or Python. However, a computer can only understand machine language which is binary—consisting of 0's and 1's.

A computer program that is written in any programming language is first converted to a machine code (binary) and stored on a storage device (mostly as a file). When this machine code is executed or read by the computer, the CPU follows the instructions mentioned in the code. A software can be of two types—system software and application software.

- System software—allows the other application software to work with the computer. It works like a middle person between the computer and the apps. As the name suggests, it's designed mainly for the computer system (hardware), and it controls the hardware using various system files and device drivers. Many OSs like Windows, macOS, Linux, and Android are some examples of system software.

  Such type of OS can have a CLI and the user needs to use various commands to operate the computer or can have a GUI that gives a graphical environment for the user to interact with the computer.
- Application software—allows the users to directly interact with the computer and perform various tasks like create text documents, play or develop games, create presentations, listen to music, draw pictures, or browse the web. Application software can be either run from a CLI or GUI, and many applications come along with an OS.

  Applications can be of many types depending on its design and use. In modern computer applications like Microsoft Office (document processing), Audacity software (audio editing), Final Cut Pro (video production), Adobe Photoshop (photo editing), and so on, are some examples of application software.

Most general-purpose computers work with an OS, which is a software that controls the hardware and enables the user to operate the computer. Most OS comes with system and applications software. OSs are developed by a software engineer/developer using a programming language like C. Every OS contains a core called kernel that interacts with the applications and the computer. It's like a middleware between the application software and the hardware, and lets the applications use various hardware resources (see figure 2.2).

The top layer is the application layer, and the user interacts with any application of an OS using the user interface. The kernel stays below the application layer and it manages the applications and handles all the requests made by the applications. The hardware stays at the bottom layer and it's the kernel that actually manages and controls the hardware. Most general-purpose computers use either a 32-bit or 64-bit OS depending on whether the CPU is 32-bit (x86 architecture) or 64-bit (x64 architecture). A 64-bit CPU works better than a 32-bit CPU and can take more processing load.

That's why using a 64-bit OS on a 64-bit CPU can take full advantage of the CPU but not when a 32-bit OS is used on a 64-bit CPU. However, a 64-bit OS can't be used on a 32-bit CPU. There are many OS available for general-purpose computers, and Microsoft Windows is most popularly used for home and business on a Windows personal computer (PC)—formerly known as IBM PC. On the other hand, macOS, also used for home and business, can

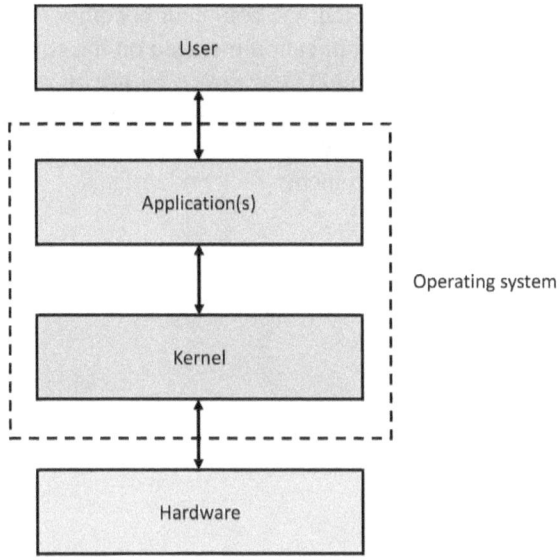

Figure 2.2 Operating System Layers. Author Created.

only be installed on an Apple PC. Either of these two OS can be installed on both types of computers—Windows can be installed on Apple PC, and macOS on Windows PC, using various hacking methods.

Microsoft Windows home version is free to download and use but most of its business versions are not free and need to be purchased. The macOS comes with the Apple PC and can be purchased from the online Apple store only. Unix is one of the oldest operating systems developed in 1969 at Bell Labs, and Apple developed macOS from the source code (the programming data) of Unix OS.

Just like Unix, another operating system called Linux was developed in 1991 by Linus Torvalds, and it's free to use or modify. Linux can be used on many x86/x64 CPUs, and there are many types of Linux distributions available for use. Among all the distributions, Ubuntu is the most popular Linux available with Windows/MacOS like GUI for use. Many OS come with applications for work and entertainment, but many other third-party applications can also be installed.

When a computer is powered on, the BIOS runs a diagnostic test called the Power-On Self-Test (POST) that checks whether the hardware components are working properly, and after a successful POST operation, the computer starts to load the OS to its memory—a process called booting. In this booting process, first the bootloader (booting instruction) is read from the storage device (like hard disk) and as per the bootloader instructions the booting of an

OS is done. After the OS is booted, the computer becomes ready to be used, and the user can then use any application installed on the computer.

There are many applications and OS available for use on a general-purpose computer and it all depends on the usage and the hardware requirements. Therefore, software should be chosen for a general-purpose computer only after knowing the system requirements.

Chapter 3

What's a CPU?

The Central Processing Unit (CPU) is like the heart of a computer, and without a CPU a computer can't work. A computer works arithmetically and all the arithmetic operations are done inside the CPU. The CPU is the actual computer that processes the input data and generates the output data. CPUs are made from transistors, and a modern CPU would contain billions of transistors that help process binary data by switching on and off to represent binary data (1s and 0s).

According to Moore's law the number of transistors in an IC doubles every two years. This is named after Gordon Moore (founder of Intel Corporation) because he perceived this theory in 1965. ICs and microprocessors used as a CPU are manufactured using a process called semiconductor device fabrication. In this process a semiconductor wafer (preferably silicon) is used to manufacture the chip containing billions of transistors, and it requires state-of-the-art manufacturing facility.

This is done with a method called photolithography, which lets the circuit to be printed on a chip die using UV light. The size of the transistors used in the chip and their distance between them is measured in micrometers (μm) or nanometer (nm). Most of the chips first manufactured in 1971 were done with a 10 μm process, but in 2022 it came down to 5 nm.

Intel 4004 was the first commercially available processor that worked as a CPU in 1971, and it was manufactured using a 10 μm fabrication process.

Intel 4004 was just a single-core CPU, but nowadays CPUs with multiple cores are getting faster and more powerful. CPUs like Intel Core i9 and AMD's Threadripper series are the most powerful CPUs used for desktop computers.

CPUs are designed for many types of computers including desktops, workstations, and servers, and their use depends on the architecture that includes the components and specifications.

COMPONENTS OF A CPU

A CPU contains the ALU, CU, cache memory, and register. However, most modern CPUs also come with integrated GPU, which improves graphics processing performance. Apart from that, an additional mathematical unit called the Floating-Point Unit (FPU) is integrated to carry out operations on floating-point numbers like addition, subtraction, multiplication, division, and square root (see figure 3.1).

- ALU—the mathematical unit of the CPU that performs all the arithmetic operations on any binary data. A basic ALU has three data buses that need two input data (A, B) and one output (Y), Opcode that specifies the ALU's operations and its bus size determines the maximum number of operations, and I/O status that tells the ALU about the status about the I/O data.
- CU—unit that controls all the components of the CPU, including the ALU, main memory, and I/O devices. It's like a director who directs the flow of

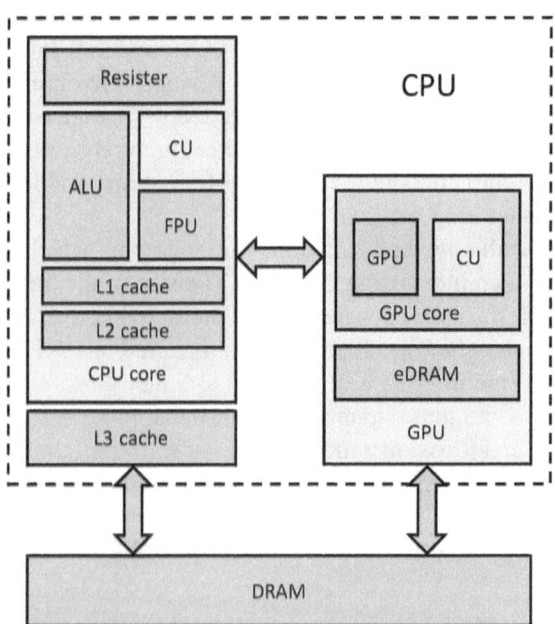

Figure 3.1 CPU Block Diagram. Author Created.

Chapter 3

What's a CPU?

The Central Processing Unit (CPU) is like the heart of a computer, and without a CPU a computer can't work. A computer works arithmetically and all the arithmetic operations are done inside the CPU. The CPU is the actual computer that processes the input data and generates the output data. CPUs are made from transistors, and a modern CPU would contain billions of transistors that help process binary data by switching on and off to represent binary data (1s and 0s).

According to Moore's law the number of transistors in an IC doubles every two years. This is named after Gordon Moore (founder of Intel Corporation) because he perceived this theory in 1965. ICs and microprocessors used as a CPU are manufactured using a process called semiconductor device fabrication. In this process a semiconductor wafer (preferably silicon) is used to manufacture the chip containing billions of transistors, and it requires state-of-the-art manufacturing facility.

This is done with a method called photolithography, which lets the circuit to be printed on a chip die using UV light. The size of the transistors used in the chip and their distance between them is measured in micrometers (μm) or nanometer (nm). Most of the chips first manufactured in 1971 were done with a 10 μm process, but in 2022 it came down to 5 nm.

Intel 4004 was the first commercially available processor that worked as a CPU in 1971, and it was manufactured using a 10 μm fabrication process.

Intel 4004 was just a single-core CPU, but nowadays CPUs with multiple cores are getting faster and more powerful. CPUs like Intel Core i9 and AMD's Threadripper series are the most powerful CPUs used for desktop computers.

CPUs are designed for many types of computers including desktops, workstations, and servers, and their use depends on the architecture that includes the components and specifications.

COMPONENTS OF A CPU

A CPU contains the ALU, CU, cache memory, and register. However, most modern CPUs also come with integrated GPU, which improves graphics processing performance. Apart from that, an additional mathematical unit called the Floating-Point Unit (FPU) is integrated to carry out operations on floating-point numbers like addition, subtraction, multiplication, division, and square root (see figure 3.1).

- ALU—the mathematical unit of the CPU that performs all the arithmetic operations on any binary data. A basic ALU has three data buses that need two input data (A, B) and one output (Y), Opcode that specifies the ALU's operations and its bus size determines the maximum number of operations, and I/O status that tells the ALU about the status about the I/O data.
- CU—unit that controls all the components of the CPU, including the ALU, main memory, and I/O devices. It's like a director who directs the flow of

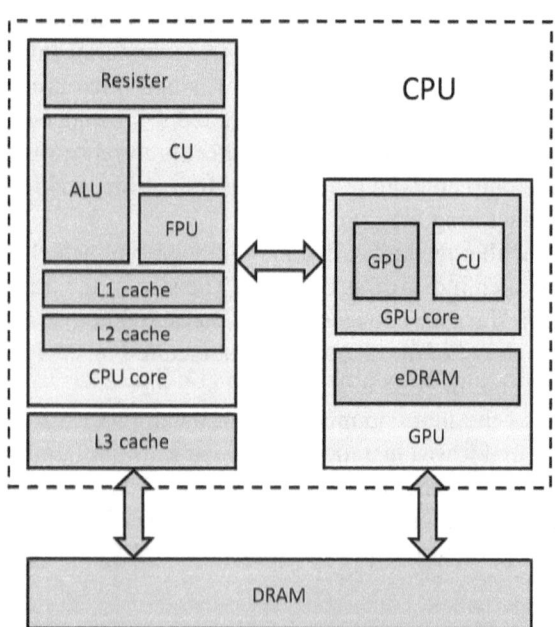

Figure 3.1 CPU Block Diagram. Author Created.

data between various devices connected to the CPU. Since the implementation of CU in CPUs the role of CU has remained the same and in modern CPUs; also CU is used to control the CPU and its connected devices.
- Cache memory—small size memory used by the CPU for storing recently processed data. Its memory stays between the main memory, which is usually the dynamic random-access memory (DRAM) and the CPU. But it's faster than the DRAM and responds to the CPU in less time. This reduces the wait time for the CPU to repeatedly read data from the memory, otherwise it would affect the system's performance during the CPU wait time.

  In most modern CPUs there exists three levels of cache memory—L1, L2, and L3. L1 is faster than L2, and L2 is faster than L3, where L3 is the outermost cache memory that stores data from the main memory.
- Resistor—memory within the CPU core that works like a cache memory but it stores unprocessed data, and also stores the addresses of the data stored in the main memory. This works like a scratch pad and the CPU uses a register frequently for processing data.
- GPU—processing unit that processes data related to graphics and it's integrated within the CPU chip in most modern microprocessor chips like Intel Core i7. It's also called an integrated graphics processor (IGP) and some motherboards also contain IGP. A GPU can also be considered as a vector processor because it can process vectors (multiple one-dimensional arrays of data used for graphics and simulation).

  Most GPUs have inbuilt memory but it can also utilize the main memory while processing data. Like a CPU, GPUs also have an operating clock speed and contain multiple cores. The GPU works along with the CPU to process the graphics data, and the processed data is sent to the video device as an output. Most modern CPUs with integrated GPU are capable of processing FHD video and can be used to play most HD video games. CPUs with integrated GPUs from AMD are called Accelerated Processing Unit (APU).
- FPU—part of a computer that is capable of processing data to carry out mathematical operations like addition, subtraction, multiplication, division, and square root on floating-point numbers. Some FPUs can process special mathematical data like transcendental functions for exponential and trigonometric calculations, but their accuracy is limited, and that's why most computers use software for such calculations.

Earlier, an independent processor called the math coprocessor was used separately to support floating-point operations, but now FPUs are integrated in the CPU as a part of its architecture, and FPU instructions are used to perform any floating-point operations. Modern CPUs from Intel and AMD use streaming SIMD Extensions (SSE) instructions instead of FPU instructions to process floating-point data.

SPECIFICATIONS OF CPU

Just like any other device or a machine that comes with specifications, a CPU also has specifications. For example, your bike or car also has specifications like mileage, power, top speed, fuel type, braking system and so on. Similarly, a CPU has specifications like frequency (clock speed), bit-length, cache size, supported memory, total cores, socket type, lithography, power consumption, and its generation. The efficiency and worthiness of a CPU depend on its specifications only.

- Frequency (clock speed)—the cycles of operation a CPU can perform in one second. It's usually measured in hertz (Hz) and 1 kHz (kilohertz) = 10^3 Hz, 1 MHz (megahertz) = 10^6 Hz, 1 GHz (gigahertz) = 10^9 Hz, and 1 THz (terahertz) = 10^{12} Hz. Earlier, CPUs used to have a clock generator that would generate the clock pulse with an operating frequency, but nowadays it's done by the clock generator on the motherboard. That's why it's possible to overclock a CPU on a mother that supports overclocking. A clock pulse with a specific frequency is used to synchronize the data flow within the CPU.

    The operating frequency of Intel 4004 chip is 740–750 kHz, and most modern desktop CPUs can now operate between 1 GHz and 4 GHz, but Intel Core i9-12900K can reach a clock speed of 5.5 GHz. In 2011, the Guinness World Record for the highest CPU clock rate was 8.42938 GHz with an overclocked AMD FX-8150 processor using liquid nitrogen for cooling, and later many other processors like AMD FX-8350 were also overclocked to reach up to 8.79 GHz of clock speed.

    The problem with overclocking is the increase in CPU temperature and with the help of effective cooling solutions like using liquid nitrogen cooling systems, it has become possible to extremely overclock the CPUs.
- Bit-length—total number of bits (data) a CPU can process at a time. It also represents the size of the resistor (memory) present in a CPU. Wider the bit-length, better gets the processing capacity. To possess one byte as data, it takes 8 bits and any 8-bit CPU can process one byte at a time only. The Intel 4004 chip is a 4-bit CPU and can process half byte (nibble) per clock cycle (1 Hz). So this 4-bit chip would require 2 Hz to process one byte of data.

    Most modern CPUs used in general-purpose computers are either 32-bit or 64-bit wide, and 64-bit CPUs are mostly used in desktops and workstations because of their better performance.

- Cache size—as this memory is used to improve the performance of the computer, the cache size of the CPU needs to be higher, and higher the cache size, smoother the computer operates (especially while playing video games). Earlier, CPUs like the Intel 486DX2 used in desktop computers during the 1990s had only L1 cache of 8 KB–16 KB.

  But nowadays, most powerful CPUs (like Intel Core i9-12900K) have 80 KB–96 KB of L1, 1.25 MB–2 MB of L2, and up to 30 MB of L3 cache memory. Among the three cache memories, L3 should be considered to be responsible for improving the system performance while playing games, and that's why most gaming CPUs of eleventh and twelfth generations have L3 cache memory of at least 12 MB.

- Supported memory—the main memory that is installed on the motherboard. This can be of many different generations and can be various sizes with different transfer speeds. The type of memory used in modern general-purpose computers is called Double Data Rate Synchronous Dynamic Random-Access Memory (DDR SDRAM), and there are five generations (as of 2022) of DDR memory used—DDR, DDR2, DDR3, DDR4, and DDR5.

  Most recent CPUs are only compatible with DDR4 and DDR5 types of memory. A DDR memory has a size measured in bytes, and a single DDR4/DDR5 module comes in sizes of 4 GB, 8 GB, 16 GB, 32 GB, and 64 GB. A CPU that can support 64 GB of maximum memory can use either a single memory module of 64 GB or four numbers of 4 GB modules or two numbers of 32 GB modules, depending on the availability of memory slots on the motherboard.

  DDRs also have a transfer rate and it's measured in mega transfers per second (MT/s). The transfer of data per second depends on the bit-length. If the bus is 64-bit (8 bytes), then 1MT/s would mean $8 \times 10^6 = 8$ MB/s (8 million bytes in one second). DDR modules are numbered in the format DDR(G)-(XXXX), where "G" represents the DDR type number and "XXXX" is the transfer speed. For example, DDR4-1600 means a DDR module of fourth generation with a transfer speed of up to 1600 MT/s.

  If a CPU supports DDR4 with a transfer speed of up to 3200 MT/s, then any DDR4 starting from DDR3-1600 to DDR3-3200 can be used, but if a CPU supports only DDR4, then a DDR5 can't be used with that CPU because DDR generations are neither forward nor backward compatible.

- Total cores—the total number of CPUs present as cores within the main CPU chip. Most previous generation CPUs like Intel 4004 and Celeron G470 had only single core, but most recent CPUs like Ryzen Threadripper PRO 5995WX and EPYC 7773X from AMD have 64 cores. The number of CPU cores improves system performance and makes a computer more powerful.

However, only operating systems and applications that support symmetric multithreading (SMT) can take advantage of multiple cores. For example, operating systems like Windows 98 won't benefit from a multicore CPU, but newer OS like Windows 11 are designed to work on multicore CPUs. Intel's hyper-threading (HT) technology makes a CPU core appear as two logical (virtual) cores for the OS, and a dual-core CPU would have four logical cores for the OS to work with.

Whenever a CPU core would stall (wait for data to process) another logical processor would carry on the remaining scheduled task. This improves the performance (up to 30 percent only) but consumes more power and generates heat. Some Intel processors belonging to the Skylake and Kaby Lake family had a bug with their implementation of HT that could cause data loss.

Later, Intel removed the HT feature in most processors of Coffee Lake except the high-end i9 processors. Intel recommends that HT feature should be disabled to avoid any hacking attacks, and unnecessarily keeping the HT feature turned on could consume extra power.

On the other hand, CPUs with more physical cores like Ryzen Threadripper PRO 5995WX improves system performance but to take advantage of such multicore CPUs the applications should be programmed to share the load efficiently across all the cores, and such programming is also very difficult to debug.

Using multicore CPUs without hyper-threading can also improve 60–80 percent system performance, but the OS should be optimized to work properly with multiple cores. CPUs with too many cores are best suitable for parallel computing, which is also used in supercomputers.

- Socket type—the type of socket on the motherboard where the CPU is installed and through the socket it gets connected to the other components. CPU sockets that are used in desktops and workstations are mostly of three types—Pin Grid Array (PGA), Land Grid Array (LGA), and Ball Grid Array (BGA).
    - PGA is the only type of CPU socket used on motherboards in the beginning, and Intel introduced it with the Coppermine core Pentium III and Celeron processors based on Socket 370. In PGA the processor with contact pins needs to be inserted in the socket on the motherboard, and such a socket is also called a zero insertion force (ZIF) socket because it needs very little force to fit the processor in the socket. AMD also introduced PGA for their line of processors and still uses it for the sockets like AM4.
    - LGA came after PGA, and most Intel processors use this type of socket. In LGA, the contact pins are on the socket and gold-plated copper pads on the processor that eliminates any accidental bending of pins, which could happen with a PGA processor. That's why LGA is better than PGA

and makes a processor more durable, and it also provides better contact with the motherboard. LGA was first introduced by Intel and later AMD also implemented it on its Ryzen Threadripper processors.
- LGA 1200 (Socket H5) and LGA 1700 (Socket V0) are the LGA sockets used by Comet Lake (10th gen) and Rocket Lake (11th-gen) Intel processors, whereas Socket sTRX4 and AM5 are used by AMD Ryzen Threadripper processors.
- BGA socket doesn't contain any pins but contains balls of solder (metal used for soldering electronic circuits). A processor can't be installed or removed from a BGA socket easily, because a processor is permanently fixed to the socket through a soldering process called Surface-Mount Technology (SMT). This is usually done during the manufacturing process and any motherboard that came with a BGA processor can't be upgraded or changed by hand.

  However, BGA can become a solution for processors with hundreds of contact points because PGA or LGA have space limitations but BGA can be implemented on smaller processors that require smaller space with more contact points. It also reduces heat and prevents unwanted inductance (disturbance in flow of electricity).
- Lithography—the process of manufacturing IC using UV light and the process is measured in nanometers (nm), which indicates the size of the transistors and their distance in the chip. Smaller the lithography process size, smaller is the chip. Most recent processors (as of 2022) are manufactured using the 7–5 nm lithography process.
- Power consumption—the amount of electrical energy consumed by the CPU under the maximum theoretical load. It's also referred to as Thermal Design Power (TDP). TDP rating of a processor indicates its maximum power consumption and heat dissipation of a processor. By knowing the TDP of a processor, a better power supply and cooling system can be selected for the computer. Most Intel processors have lower TDP values than the AMD processors because most AMD processors come with too many cores.
- CPU generation—just like people born and living at about the same time belongs to one generation—processors that are manufactured with similar technologies during a specific timeline fall into the same generation. Intel uses numbers to indicate a processor's generation, and starting from first (2010) to twelfth generation (2022) the first two numbers of the Intel core series processors indicate its generation. For example the processor i7-7700K belongs to the seventh generation because the number "7" right after the dash indicates seventh generation.

  AMD processors also have generations and most desktop processors of the Ryzen series use the second number to indicate its generation. For example the AMD processor Ryzen 9 3900XT belongs to the third generation because the number "3" after the Ryzen series number "9" indicates its generation.

By knowing the processor generation, it's possible to select the right processor for the motherboard. Some motherboards can support recent and previous generation processors and the details can be found in the datasheet.

Apart from the components and the specification, there are many other things that are responsible for the CPU's performance like the architecture and design. Architecture like x86-64 is based on the Intel 8086 processor and that's why the number "86" is used after "x." Whereas, an Arm64 architecture means any processor based on the architecture of the 64-bit ARM processor—commonly used in smartphones.

Design of a processor is based on the type of instruction set used, and usually there are five types of instruction sets—Complex Instruction Set Computer (CISC), Reduced Instruction Set Architecture (RISC), Minimal Instruction Set Computer (MISC), Very Long Instruction Word (VLIW), and Explicitly Parallel Instruction Computing (EPIC). Most x86-64 architectures have CISC and most desktop CPUs are CISC. ARM processors used in portable devices like laptops and smartphones have RISC design.

MISC was used in a processor called Transputer in the 1980s that used parallel computing. VLIW was used in the processors like Transmeta Crusoe developed by Transmeta in 2000 and Elbus 2000 developed by Moscow Center of SPARC Technologies (MCST) in 2008. EPIC was used on Intel IA-64 architecture processors like Itanium in 2001, but it didn't perform well in the long run and was discontinued in 2019. Among all these, the x86-64 and Arm64 designs are currently used in most processors.

Therefore, a CPU with robust components and good specifications can make a computer work perfectly. However, other components like memory and I/O devices that are externally connected to the CPU are also responsible for the overall performance of a computer.

Chapter 4

What's a GPU?

Graphics processing unit (GPU) is a processor like the CPU of a computer, capable of processing graphics/video data at a particular frame rate (frames of images per second). With the help of a GPU it is possible to either process digitally stored image/video data to get it displayed on a video output device like the computer monitor (LCD/LED display) or digitally generate any graphics using software. A GPU is used in a computer to view image files, play videos, and run computer video games.

It's also used to run any operating system (OS) or any application that has a GUI. Even though a good GPU is required to run any software or games that generate graphics, at least a low-end GPU should be present in a computer to get the display on a computer monitor, in order to operate the computer.

If GPU is not used, then the CPU can't do graphics processing alone and that's why even the arcade video game boards have used specialized graphics circuits during the 1970s. The Atari 8-bit computers of 1979 used a separate video processing chip called Alphanumeric Television Interface Controller (ANTIC). In the 1970s the term GPU originally stood for *graphics processor unit* but Sony later termed it as *graphics processing unit* in 1994 when Toshiba designed the Sony GPU for Sony's PlayStation consoles. This term became popular when Nvidia marketed the world's first GPU called the GeForce256.

Just like the chip of a CPU, a GPU chip is also manufactured by a fabrication process using photolithography, and its process is measured in nanometers (nm). The Nvidia GeForce256 GPU chip was manufactured by Taiwan Semiconductor Manufacturing Company Limited (TSMC) with a 220 nm process.

The GeForce256 was released in the market with two versions—one with Synchronous Dynamic Random-Access Memory (SSR SDRAM) and another with DDR SDRAM. SSR SDRAM is predecessor to DDR SDRAM and is much slower.

A GPU also has clock speed (frequency) measured in hertz (Hz), but unlike a CPU, it has inbuilt memory (different from registers and cache). The memory is used to process data and it's similar to how a CPU uses the main memory for data processing. The Nvidia GeForce256 had a clock frequency of 120 MHz and a memory of 32 MB (SSR)/64 MB (DDR). Nowadays, GPUs like the Nvidia GeForce RTX 3090 Ti have a clock speed of 1.67–1.86 GHz with a memory of 24 GB.

GPUs have a higher number of cores when compared to CPUs and can have thousands of GPU cores. In order to run video games and certain graphics processing applications, a GPU should support 2D/3D graphics Application Programming Interface (API) like Direct 3D (part of DirectX) of Microsoft, OpenCL, OpenGL, and Vulkan. If a GPU fails to support any of these APIs, then a software or game using that API won't run properly or may not run at all. That's why before choosing the GPU for a computer ensure that it supports all of the APIs.

COMPONENTS OF A GPU

A GPU like the ones from Nvidia works mainly using components like CUDA/GPU cores, memory controller, cache memory, and graphics memory (GDDR SDRAM). (See figure 4.1.)

- CUDA/GPU cores—multiple cores used in a GPU that help in processing graphics data. CUDA is actually an API that allows parallel computing on a GPU. As the virtual instruction set of Nvidia GPU cores can be accessed using CUDA for general-purpose graphics processing, each core is called a CUDA core. This makes it easier for applications and games to make use of the GPU using parallel computing, and CUDA also supports C/C++ programming language.

    The CUDA cores of a GPU are mainly responsible for processing or generating computer graphics. Generally the multiple cores of a GPU are called GPU cores, and can support parallel computing through other APIs also.
- Memory controller—the unit that controls the memory of the GPU. Unlike the CU of a CPU that controls all the components, the memory controller controls only the memory (eDRAM) of the GPU.

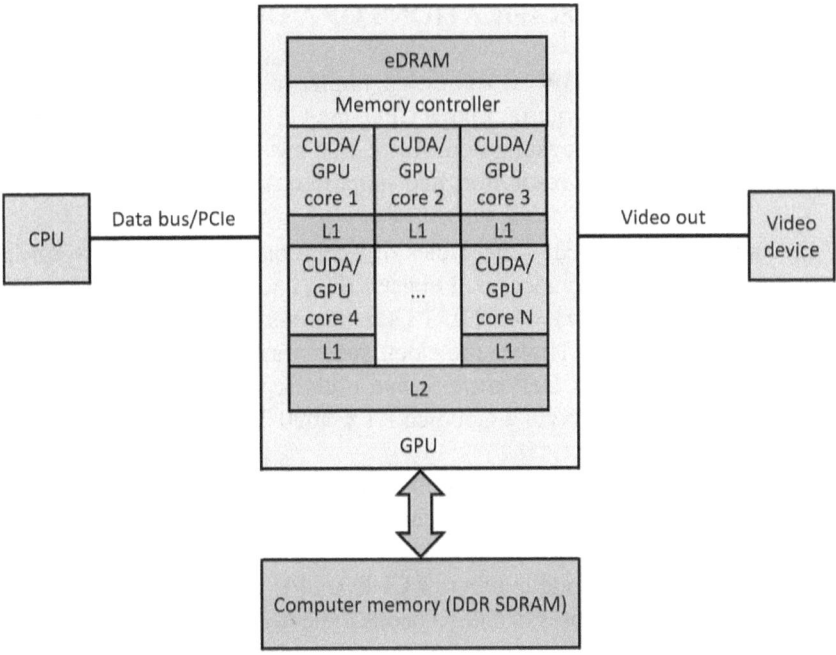

Figure 4.1 GPU Block Diagram. Author Created.

- Cache memory—a small size memory that is faster than the main memory, used to improve the performance. GPU can have L1 and L2 cache memory but L3 may not be present. L1 is used in every core but L2 is shared by all the cores.
- Graphics memory—a memory similar to the main memory of the computer that is used by the GPU to store processed data. It's usually called the embedded dynamic random-access memory (**eDRAM**) and the size of the memory is smaller than the main memory (**DDR SDRAM**) of the computer. It's like a dedicated memory used only for storing processed graphics data and that's why the size of this memory can be smaller than the main memory of the computer.

However, the GPU may use the main memory of the computer if the GPU's memory gets full or not enough to process the graphics data in some situations. In today's modern graphics cards like GeForce RTX 3090 Ti has memory a size of 24 GB. But it's not always necessary to use a GPU with such a high memory size, and for playing most HD games or using creative applications, a graphics card with a memory between 4 and 8 GB could be sufficient.

SPECIFICATIONS OF A GPU

The specifications of a GPU is somewhat similar to the CPU, but includes a few different technical criteria. Like a CPU, the GPU also has a clock speed, total cores, and power consumption. But it also has other specifications like memory size, maximum resolution, and supported API.

- Frequency (clock speed)—the cycles of operation a GPU can perform in one second. It's usually measured in hertz (Hz) and 1 kHz (kilohertz) = 10^3 Hz, 1 MHz (megahertz) = 10^6 Hz, 1 GHz (gigahertz) = 10^9 Hz, and 1 THz (terahertz) = 10^{12} Hz. Higher the clock speed, faster is the GPU. Earlier, GPUs like the Nvidia GeForce256, had a clock speed of 120 MHz, but nowadays GPUs like Nvidia GeForce RTX 3090 Ti have a clock speed of 1.67–1.86 GHz.

  Just like CPU overclocking, a GPU can also be overclocked using software tools, and GPU overclocking can improve graphics processing (especially while playing video games).
- Total cores—it's the total number of CUDA/GPU cores present in a GPU. The total number of cores in a GPU needs to be higher than a CPU because graphics processing requires too much processing to be done at the same time, which is like parallel computing. Therefore, sharing the workload among different GPU cores improves performance.
- Power consumption—this specification is similar to a CPU because like a CPU, a GPU also consumes power and it's also measured in watts (W). Higher the wattage of a GPU, greater the amount of heat produced during full load. That's why a GPU with higher wattage value would require more cooling fans for better cooling. High-end GPUs used in gaming desktops and workstations require more than 300 W power to operate properly.
- GPU memory (eDRAM)—is the main memory (apart from the cache memory and the main memory of the computer) that is used by the GPU cores to store processed data for generating graphics as video output. The GPU memory size is one important specification that tells how powerful it is, and higher the memory size, better is its performance while playing HD video games like *Assassin's Creed: Odyssey*. Therefore, a GPU with a memory of 8 GB is best for running most HD games, and also for running any 3D simulation software for productivity.
- Maximum resolution—it's the highest resolution supported by the GPU. Resolution means the number of square-like dots (pixels) represented on a computer display like an LCD to show an image. The display resolution is defined by the total number of pixels present both horizontally and vertically. For example, an HD display of 720p would require 1280 pixels

horizontally and 720 vertically on a display to represent an HD image of 720 p (1280 × 720).

Similarly, a full HD (FHD) image requires 1080p (1920 × 1080) resolution on the display screen. That's why the GPU must generate the video data at 1080p in order to view the video on a 1080p LCD display. If the maximum resolution of a GPU is 1980 × 1080, that means it can only generate video upto 1080p (FHD), but it won't be able to generate or process video at 4K resolution, and using a 4K display with such a GPU will not make any difference.

That's why the maximum resolution of the GPU should be checked before using any display device.

- Supported API—it's the specification that tells which APIs are supported by a GPU. API is used by video games and many GUI software to render graphics, and it depends on the game or the software that requires a particular API to use the GPU. There are many APIs like Direct 3D, OpenGL, OpenCL, and Vulkan that are used by many video games and applications to render graphics using a GPU, but if the GPU doesn't support any of these APIs, then the game or application that needs the API to access the GPU won't run properly or may not run at all.

That's why a GPU that supports all of these APIs will be better.

GPUs from Nvidia and AMD come in various types of series. And the performance of the GPU can be determined from the series type. Nvidia GPUs are either of the GTX or RXT type and the RTX types are better than the GTX, but the RTX type comes in RTX 20 and RTX 30 series, and RTX 30 series GPUs are much more powerful than the RTX 20 series. However, higher the series number, better is the performance, but higher the cost of the GPU. AMD GPUs come as Radeon series of RX and PRO versions.

The RX series is for desktops and the number after the suffix "RX" indicates its value and precedence. For example, the difference between Radeon RX 6800 and Radeon RX 6950 XT is that the GPU of RX 6800 is less powerful than RX 6950 because the number "6950" is greater than "6800." It also indicates that the GPU of RX 6950 is a successor to RX 6800, and could perform better being an updated one. Radeon PRO series GPUs are for workstations and use the letter "W" followed by a number to indicate the series. Similarly, a Radeon PRO W6800 can outperform the Radeon PRO W6600 Graphics because of having a higher series number.

A GPU can be external or integrated. The external GPU can be either installed on a motherboard as a graphics card through the Peripheral Component Interconnect Express (PCIe) slot or it might be inbuilt on the motherboard. Integrated GPUs are those that come with the processor of the CPU. Most recent processors of Intel and AMD come with integrated GPUs.

The GPU that comes integrated with an Intel processor is called UHD or Iris Xe graphics depending on the processor number. But in the case of AMD, the processor of the CPU that comes with an integrated GPU is called Accelerated Processing Unit (APU), but not all AMD processors are APUs.

Therefore, a GPU with robust components and good specifications can generate graphics nicely. However, other components like CPU, main memory, and display devices that are externally connected to the GPU are also responsible for the performance of a computer, while playing video games or using any application that requires immersive graphics processing.

Chapter 5

What's Computer Memory?

A computer memory is an electronic component (usually a chip) that holds data, which is either to be processed or already processed by the CPU. It's like a notepad to the computer, and without the memory a computer can't work. There can be two types of memory—volatile and non-volatile. A volatile memory is any electronic component or device (connected to the computer) that can store data temporarily because the data gets wiped out after the computer is powered off.

On the other hand, a non-volatile memory is the one that can store data (permanently), even after the computer is powered off. That's why a volatile storage is used in a computer only to process data and to store the process data temporarily before saving it on a non-volatile memory. RAM is a volatile memory chip used by the computer as the main memory for data processing. There are various types of RAM used in a computer, but the most commonly used one is called double data rate synchronous dynamic RAM (DDR SDRAM).

The chip that holds the BIOS settings is called the Complementary Metal-Oxide-Semiconductor (CMOS) is a volatile memory, and it needs power to keep the BIOS settings. That's why a coin battery like the CR2032 is used as a backup power to keep the BIOS settings in the CMOS chip even after the computer is powered off. It is also called as non-volatile RAM (VRAM) because the chips can store the BIOS settings even when the computer is powered off, but that is done with the help of the CMOS battery only. This chip is still a volatile memory and can lose data if the battery is removed.

Non-volatile memory is usually a storage device (like hard disk) connected to the computer that can store the processed data permanently. A hard disk is a magnetic disk that can store data with the help of an electromagnetic recording method similar to that of audio tapes. But a hard disk uses digital data to store on the magnetic disk, whereas an audio tape recorder uses analog signals to record on the tape. A hard disk is the most commonly used non-volatile storage used for storing large amounts of data.

There are some chips like Programmable Read-Only Memory (PROM) and Electrically Erasable Read-Only Memory (EEPROM) that can store data permanently, even after the power is switched off. PROM chips are only One Time Programmable (OTP) and whatever data is written to the chip cannot be changed later, but a EEPROM chip can be reused to store data permanently. Many electronic gadgets like media players, video game consoles, smart TVs, and so on, come with such chips to hold the firmware permanently but can later be updated also.

The chip that holds the firmware of BIOS is a EEPROM and doesn't require any power to hold the data permanently.

Flash memory is the most effective type of non-volatile memory used in modern computers to store data permanently. Flash memory is either NOR or NAND type (similar to the NOR/NAND logic gates), and NAND is the most common type of flash memory found in memory cards, USB flash drives, solid-state drives, and so on.

A computer memory (both volatile and non-volatile) can also be considered as a primary and a secondary memory. Usually, the RAM is considered as the volatile primary memory, and the other non-volatile memory like the hard disk is considered as the secondary memory.

RANDOM-ACCESS MEMORY

The first RAM was the Williams tube—a CRT that stored data as electrically charged spots on the screen. It was invented by Freddie Williams and Tom Kilburn in 1957, just before the invention of IC. But such CRT-based RAM was not a practical way to store data from a computer. That's why magnetic cores were used in computers to work as RAM, but they were extremely slow. Later, in 1964, John Schmidt at Fairchild Semiconductor developed a type of RAM called the Metal-Oxide-Semiconductor RAM (MOS RAM).

The performance of MOS RAM was better than magnetic cores and also consumed less power. In 1968 MOS IC was developed by Federico Faggin at Fairchild, which led to the production of MOS memory chips.

The static RAM (SRAM) was invented by Robert H. Norman at Fairchild Semiconductor in 1963, and MOS SRAM by John Schmidt at Fairchild in 1964. In 1965, the commercial use of the MOS SRAM chip started and IBM introduced the SP95 memory chip for the System/360 Model 95. The dynamic static RAM (DRAM), which used capacitors and transistors to store data, was used in Toshiba's Toscal BC-1411 electronic calculator in 1965.

Later in 1966, Dr. Robert H. Dennard at the IBM Thomas J. Watson Research Center realized that MOS capacitors can be used to store binary data along with a MOS transistor, which led to the development of single-transistor DRAM memory cells. The first commercial DRAM memory chip was Intel 1103, which was manufactured on an 8 μm with a storage capacity of 1 kbit, and it was released in 1970.

But the ultimate game changer was Samsung Electronics, when SDRAM was developed by Samsung, and the first commercial SDRAM chip was the Samsung KM48SL2000, which had a capacity of 16 Mbit, and it was introduced in 1992. Later, DDR SDRAM was introduced by Samsung, and a 64-bit DDR SDRAM chip was released in June 1998.

The graphics DDR synchronous graphics RAM (GDDR SGRAM), a high-speed memory used in GPU, was also introduced by Samsung, and a 16-bit GDDR chip was released in the same year. Since then several generations of DDR SDRAM starting from DDR2 to DDR5 dominated the computer industry.

Components of a RAM

A RAM chip consists mainly of memory banks containing transistors and capacitors. It is connected with the memory controller and the CPU through control and data bus. The memory controller is connected with the CPU and takes a read/write request that is sent to the RAM chips through the control bus, and data is read or written to the RAM through the data bus (see figure 5.1).

- Memory controller—it controls the RAM chip using the control bus. The CPU sends a read/write request to the memory controller and the memory

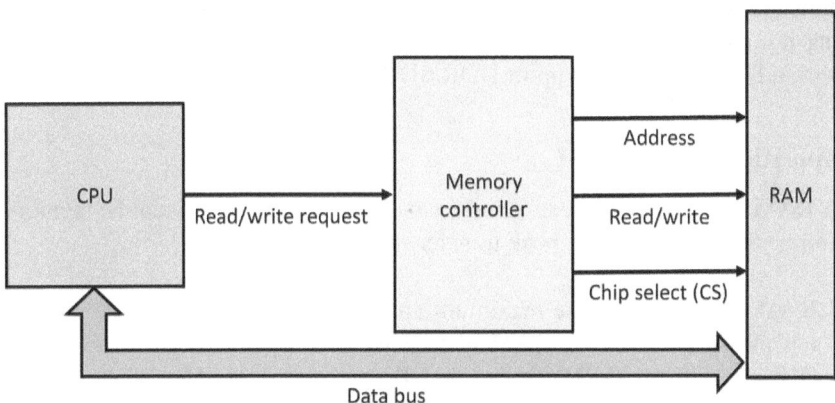

Figure 5.1 Block Diagram of RAM. Author Created.

controller takes the request and makes the RAM chip to either read the stored data at a particular address or write data to a new address. The requests are controlled by the memory controller by first selecting the chip using the chip select (CS) control and then sending the address location to the RAM.

The mode of operation (read/write) is done through the read/write selection request. The clock speed is also synchronized by the memory controller with the CPU. The requested read/write data is transferred to the CPU from the RAM through the data bus.

- Memory Bus—the connections through which data flows between the RAM, memory controller, and the CPU. There is a control bus that is used to transfer control requests between the memory controller and the RAM, and the data bus is used to transfer the data between the CPU and the RAM.
- Memory banks—the chips on the RAM that store data and the number of chips present on a RAM depend on its size and generation. The memory banks have physical addresses that can be accessed using the memory controller, and such addresses are in machine code but can be mapped to logical addresses using software. Any data can be written or read from the memory banks using their physical addresses only.

There are various types of RAM modules like Single In-line Pin Package (SIPP), Single In-line Memory Module (SIMM), Dual In-line Memory Module (DIMM), Rambus In-line Memory Module (RIMM), Small Outline DIMM (SO-DIMM), and Small Outline RIMM (SO-RIMM).

Among all these, only the DIMM and SO-DIMM modules are commonly used nowadays. The DIMM modules are used in desktops, whereas SO-DIMMs are used in laptops. DIMM modules came after the SIMM, and the difference between them is that a SIMM module had contact pins only on one side that was a 32-bit bus. On the other hand, a DIMM has on both sides making it a 64-bit bus and that's why it's called double in-line. A motherboard with a DIMM slot can support DDR SDRAM but generations can differ.

Specifications of RAM

A RAM can have specifications like capacity, generation, standards, memory and bus clock frequency, peak transfer rate, and voltage.

- RAM capacity—it's the maximum amount of data that can be stored and it's measured in bytes. RAM of various generations can be of kilobytes (KB), megabytes (MB), gigabytes (GB), and terabytes (TB), where 1 TB = 10^3 GB = 10^6 MB = 10^9 KB = 10^{12} bytes. Nowadays, DDR4 SDRAM comes as 4 GB, 8 GB, 16 GB, 32 GB, 64 GB, and 128 GB modules.

Using a RAM with lower capacity value will work with a computer but exceeding the maximum supported capacity won't work. Modules with different capacities should not be installed on a dual-channel motherboard because that will force the RAM to run in single-channel mode and can reduce the performance.
- RAM generation—just like people born and living at about the same time belong to one generation, RAM chips that are manufactured with similar technologies during a specific timeline fall into the same generation. This concept is applicable especially for the DDR SDRAM and a number right after "DDR" is used to denote the generation. For example, DDR2 means second generation, DDR3 means third generation, DDR4 means fourth generation, and DDR5 means fifth generation.

  As of 2022, DDR4 and DDR5 are the currently used RAM generations. A motherboard can support RAM belonging to the same generation only, which means that if a motherboard datasheet says that it supports DDR4, then neither DDR5 nor any other generations below DDR4 can be installed. A motherboard can be backward compatible with processor generations but not with RAM, and higher the generation of RAM, better is its performance.
- Standards of RAM—a RAM of any generation has standards that are numbered according to its transfer rate. This numbering convention is done especially for the SDRAM and a number is mentioned after the generation number separated by a dash, indicating its transfer rate. For example, DDR4-1600 is a standard that indicates the SDRAM belonging to fourth generation has a transfer rate of 1600 MT/s—can reach a peak of 1600 million data transfer in one second ($8 \times 1600 = 12{,}800 = 12.8$ GB/s).

  Higher the standard number, better is its transfer speed, but if a motherboard doesn't support that speed, then it would run at a slower speed (matching the supported speed).
- Memory and bus clock frequency—these are the two frequencies at which the RAM and the bus works together. Usually both the frequencies remain the same and it's controlled by the memory controller. Higher the frequency, higher is the transfer rate, but this specification may not always indicate the peak data transfer rate. That's why the number of the RAM standard is the best way to understand the maximum peak data transfer rate of an SDRAM.
- Peak transfer rate—it's the maximum amount of data transfer of a RAM in one second. This is usually measured in millions of transfers in one second (MT/s). The RAM standard number of a DDR SDRAM also indicates its maximum peak data transfer. For example, a DDR4-3200 SDRAM can reach up to 3200 MT/s of maximum data transfer rate. Peak transfer rate also indicates how fast a RAM can work, and using a RAM with higher peak transfer rate can improve the performance of a computer.

- RAM voltage—it's the voltage at which the RAM chip can operate. Voltage higher than the rated one can damage the RAM chip and also using it at a lower voltage can cause malfunction. Therefore, a RAM chip should receive the right voltage in order to operate properly. The one incredible fact about the DDR SDRAM is that the operating voltage is getting lower by generation.

In the beginning, the first generation (DDR SDRAM) used to operate at 2.5 V but the second generation (DDR2 SDRAM) was operating at 1.8 V. Similarly, DDR3 SDRAM used to operate at 1.5 V, DDR4 SDRAM at 1.2 V, and now the DDR5 SDRAM takes 1.1 V only. In case of RAM voltage, lower the voltage consumption, more energy efficient.

HARD DISK DRIVE (HDD)

Besides the RAM, a non-volatile storage was also used with a computer from the beginning. Punched cards were mostly used since 1920 for storing data permanently. Data was stored on a punched card by punching holes (using electromechanical punching machines) representing binary data. But punched cards were replaced by other storage like magnetic tapes during the 1960s. In the late 1980s punch cards gradually became obsolete after the use of magnetic disks for storing data.

IBM 350 was the first commercial digital disk storage device in 1956 and it was a part of the IBM 305 RAMAC computing system with a storage capacity of 3.75 MB. Later, many types of disk storage devices were released by IBM. During the 1990s many manufacturers like Seagate and Western Digital launched various types of hard disks for computers in the market.

Components of an HDD

An HDD contains components including the logic board, actuator, actuator arm, head, platter, and spindle. The logic board and the head are the two important components that make the HDD work (see figure 5.2).

- Logic board—this circuit board of the HDD controls the magnetic disk and the head. It also works as an interface between the computer and the disk. The logic board lets the computer be connected to the disk using a SATA connection. The logic board is powered by the power supply of the computer and needs 12V and 5V. The chips on the logic board can take the read/write request from the computer and control the magnetic disks to read or write data electromagnetically.

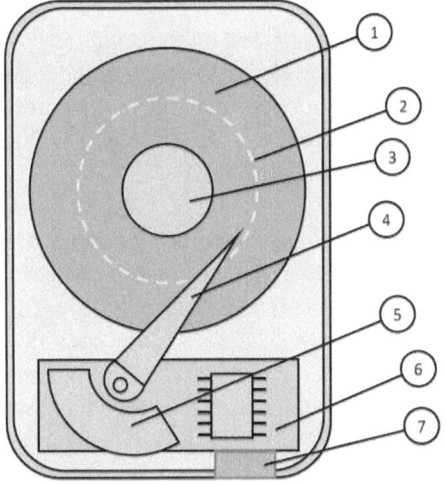
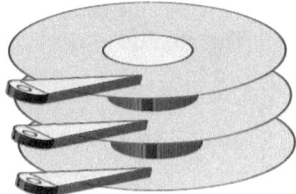

Group of heads and platters

1) Magnetic disk
2) Track with sectors
3) Spindle
4) Actuator arm with head
5) Actuator
6) Logic board
7) Interface connector (SATA)

Figure 5.2 **Components of HDD.** Author Created.

- Actuator—it's a mechanical part of the HDD that moves the actuator arm to position the read/write head of the HDD. The movement of the actuator is controlled by the logic board using electromagnets.
- Actuator arm—this is a long metallic arm that has a tiny read/write head at one end. The head is kept pointing down so that the electromagnetic flux would be recorded on the magnetic disk but the head is kept close to the surface of the disk in such a way that it would neither touch it nor cause any problem while reading or writing data on the disk. The arm can take the head freely to any location on the disk seamlessly to perform read/write operations.
- Read/write head—it's the electromagnetic head (similar to the one present in an audio tape recorder) that can read/write data on the magnetic disk. This head is connected to the logic board and while reading data, the electromagnetic signals are sent to the logic board, where it's interpreted to binary data and sent to the computer for processing. During the write operation, the binary data is converted to electromagnetic signals and are recorded on the disk.
- Platter—the disk that stores the data electromagnetically. It's usually made from glass or aluminum and coated with metallic alloy. The hard disk comes in two platter sizes—3.5 inch and 2.5 inch. These two sizes are also used to denote the size of an HDD.
- Spindle—it's the round cylindrical rotating part where the platter is attached. The spindle is usually connected to an electric motor that rotates at high speed. The rotation of the spindle also makes the magnetic disk

rotate at the same speed. The read/write operation of the head is only possible while the platter is rotating and this is controlled by the spindle.

The spindle motor is also controlled by the logic board and maintains its speed in revolutions per minute (RPM) during every read/write operation. But the maximum RPM of the spindle is fixed and an HDD with 7,200 RPM can have better response than the one with 5,400 RPM.

An HDD can have more than one platter, actuator arm, and head, arranged in a way that they are placed on top of one another so that all of them can work at the same time. This arrangement is done in order to increase the capacity of the HDD by adding more platters and heads. The read/write operation of the heads is done by moving the head across the surface of the platters from edge to edge, and data is written in thin concentric bands, which are called tracks. Each track is divided into sectors and each sector consists of 512 bytes.

Data is stored and read randomly on an HDD and that's why it's faster than a tape drive, which can read/write data sequentially. However, the random storage of data on an HDD can cause fragmentation of data over the platters and can increase the booting time of an operating system, and can also increase the loading time of many applications. To resolve the fragmentation problem of an HDD, a defragmentation software like PerfectDisk can be used.

Specifications of HDD

A magnetic HDD comes in two sizes—2.5 inch and 3.5 inch—but has similar specification criteria like capacity, RPM, interface, and transfer rate.

- Capacity—it's the maximum amount of data (in bytes) an HDD can store. Earlier the 3.5-inch HDDs were capable of storing large amounts of data up to several gigabytes (GB), but now the 2.5-inch drives can also store data up to 5 TB.
- RPM—it's the maximum revolutions per minute (RPM) of the disk, and both 3.5-inch and 2.5-inch HDDs have different RPM values. But an HDD can have either a RPM of 5,400 or 7,200 of both 2.5-inch and 3.5-inch size of different capacities. Higher the RPM, better is the hard drive performance. However, a 3.5-inch drive can have a RPM of 7,200 with a capacity of 2 TB, and a 2.5 inch can have that RPM only up to 1 TB. Some drives have RPM of 10,000 and 15,000 also but are not commonly used.
- Interface—it's a connectivity standard through which an HDD is connected to a computer. Earlier the interface was called parallel ATA (PATA) but nowadays it's serial ATA (SATA). SATA has mainly three versions—SATA

I, SATA II, and SATA III. SATA I has the slowest transfer rate of 1.5 Gb/s, SATA II has 3 Gb/s, and SATA III gives double transfer rate of 6 Gb/s. If the HDD supports SATA III but the mother supports only SATA II, then the transfer rate will be at 3 Gb/s only.

That's why both the HDD and the motherboard should support the same SATA versions to get the maximum transfer rate.

- Transfer rate—it's the maximum data transfer (in bytes) an HDD can do in one second. Even though the transfer rate of SATA III is 6Gb/s, an HDD that supports SATA 3.0 won't practically transfer data at 6Gb/s. That's why the transfer rate of an HDD is different and always lower than the interface. To know the maximum transfer rate of an HDD, it's better to refer to the manufacturer's datasheet. Most magnetic HDDs can give a transfer rate up to 200 Mb/s. However, the transfer rate may be higher for drives with higher RPM.

SOLID-STATE DRIVE (SSD)

An SSD is a type of non-volatile storage device that uses flash memory chips (usually NAND chips) to store data. Unlike an HDD that uses mechanical parts and magnetic disk to store data, an SSD doesn't have any mechanical moving parts. That's why SSDs are less prone to accidental damage from shocks and g-forces. Because of being more durable than the HDDs, SSDs are mostly used in mobile computing devices like laptops and smartphones also. Moreover, SSDs don't cause much latency because of not having any mechanical read/write system, and don't suffer from fragmentation.

SSDs used for internal storage usually come as a 2.5-inch drive with a SATA III interface, but the M.2 SSD is a memory module that fits directly in the M.2 slot of the motherboard. M.2 slot could either support SATA III interface or peripheral component interconnect express (PCIe) using the non-volatile memory express (NVMe). NVMe M.2 SSD is faster than SATA III and can reach a data transfer rate of 32 Gb/s.

The first SSD-like HDD was the StorageTek STC 4305, developed in 1978, that used DRAM, and cost about $400,000 for 45 MB capacity. However, the first flash-based SSD was invented by Fujio Masuoka at Toshiba in 1980. Later, SanDisk Corporation patented flash-based SSD in 1989 and shipped the first commercial SSD in 1991. It was a 20 MB SSD that was sold to IBM for $1,000, which was used in the Thinkpad laptops.

Since then many other companies have started manufacturing SSDs and with the improvement of flash memory and reduction in their prices, SSDs use in computers have increased dramatically. Even most HDD manufacturers are restoring their business to the manufacturing of SDDs.

Components of an SSD

SSDs work almost similarly to the RAM and have no mechanical parts like the magnetic HDD. An SSD HDD comes in 2.5-inch size as a replacement for the magnetic HDD of the same size. It usually consists of a host interface, SSD controller, memory interface, and flash memory chips (see figure 5.3).

- Host interface—it's the interface through which the SSD can transfer data to the computer, and its usually SATA for a 2.5-inch standard SSD or PCIe for M.2 SSD. The host interface builds connections between the computer and the SSD.
- SSD controller—it controls all the components of the SSD and performs the read/write operations on the SSD. It's like a CPU of the SSD that processes the read/write request of the computer and it can also be considered as the heart of the computer.
- Memory interface—it's the interface that lets the SSD controller access the flash memory chips. This is similar to the memory controller of the RAM. The memory interface can help access the memory locations of the SSD to read/write data.
- Flash memory chips—the chips that actually store data on the SSD and are usually NAND flash chips. On an SSD drive there can be several flash chips to store data and it depends on the capacity of the SSD. The response time and transfer rate of the SSD also depends on the performance of memory chips.

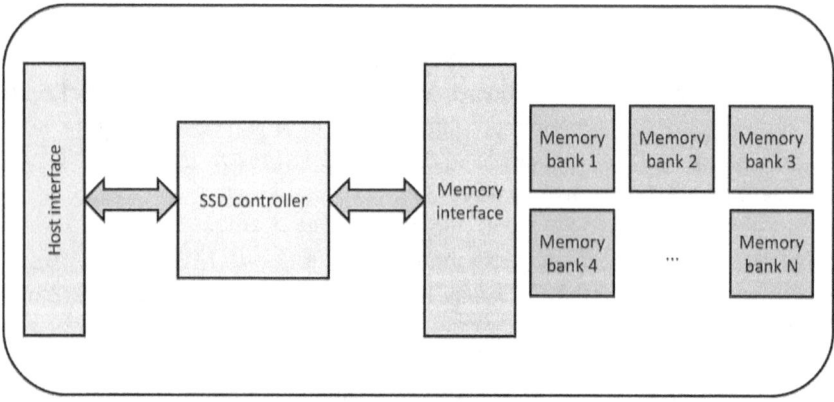

Figure 5.3 **Block Diagram of SSD.** Author Created.

Specifications of SSD

Like the HDDs, SSDs also have similar specifications like Capacity and Interface, but the data transfer rate is specified as sequential read/write speeds. As the SSDs don't have any rotating disk, it doesn't have any RPM like the HDD.

- Capacity—it's the maximum amount of data (in bytes) an SSD can store. Earlier SSDs were not capable of storing large amounts of data, but now a 2.5 inch SSD can store up to 4 TB. Higher the capacity, more expensive is the SSD, and that's why SSDs are not suitable for storing large amounts of data cheaply, but can be used to install the operating system for fast boot-ups.
- Interface—it's a connectivity standard through which an SSD is connected to a computer. As SSDs came after the HDDs, most 2.5 inch SSDs support the SATA III interface. M.2 supports SATA III as well as PCIe (via NVMe), but M.2 slot on the motherboard should support either of the interfaces.
- Sequential read/write speed—it's the read and write speed of the SSD while reading or writing data sequentially. The sequential speed while reading data is often higher than the writing speed. The sequential read/write speed of an SSD is included in the datasheet of most manufacturers as megabytes per second (MB/s) and that also signifies its average data transfer rate. Most 2.6 inch SSDs from Western Digital have a sequential read of 545 MB/s or more.

HDDs are manufactured using low-level formatting at the factory to mark each and every sector present on the platters, but this is not done for SSDs because they are not mechanical—instead, zero filling method is used to erase the memory blocks. Both HDDs and SSDs are made ready for use by high-level formatting with file systems like FAT32 or NTFS.

Apart from the internal memory, there are other storage devices like Network-Attached Storage (NAS) and online storage (cloud drive) that can be used with a computer to store data, but the transfer speed won't be as fast as the internal memory. Moreover, using online storage services like Google Drive may not always be reliable because if anything goes against the policy then the account may be suspended. Such online storage also requires a good internet bandwidth to store large files in less time.

Whether it's volatile or non-volatile, memory is an important part of a computer system and selecting the right type of memory can make a computer work perfectly.

Chapter 6

What Are Computer I/O Devices?

The Input/Output (I/O) devices of a computer are the ones that let the user enter data for processing and also provide the processed data to the user. With the help of input devices one can operate and control a computer, and the output devices can also give the status of a job while operating it. Although a storage device like a hard drive can let the computer read/write data, it can't be considered to be an I/O device for the user because the user can't use a storage device for operating the computer.

Devices like a keyboard, mouse, joystick, and so on, are some examples of input devices that can be used by a user to operate and control a computer. Similarly, a LCD/LED display is a standard output device used to view the user interface like the CLI and GUI while operating a computer.

INPUT DEVICES

There are many types of input devices that can be used with a computer, but the most commonly used are keyboard, mouse, and joystick.

Keyboard

A keyboard is used as an input device with a computer to enter data by the user. A keyboard consists of keys representing alphabets, numbers, special characters, and control keys. Since the invention of teleprinters (electromechanical devices that can transmit typed messages) in the 1830s, there have been many developments of a computer keyboard. However, the model M keyboards manufactured by IBM in 1985 was the first computer keyboard that was similar to the modern keyboard with QWERTY layout.

A standard US keyboard contains 101 keys, but nowadays a modern Windows keyboard has 104 keys or more.

Initially most desktop keyboards used a PS/2 port for connectivity but later USB port was used. Nowadays most keyboards have become wireless and can be connected using Wi-Fi or Bluetooth connection. Most keyboards for desktop computers have three light indicators for the "Num Lock," "Caps Lock," and "Scroll Lock" keys, which indicate whether any of these keys are ON or OFF. But such indicators are not seen on portable wireless keyboards or on the keyboards of laptops.

Most keyboards for desktop computers can be either a basic one or with multimedia features. A normal standard Windows keyboard can be used with a desktop computer for any home or office use. However, a multimedia keyboard can be used for gaming and for running multimedia applications like a media player. Besides the normal 104 keys, a multimedia keyboard can have additional keys for controlling a media player or playing video games.

Many gaming keyboards also come with a backlit function that illuminates the keys to use in dark environments. Some wireless keyboards also come with a touchpad that can be used like a pointing device.

Components of a Keyboard

A keyboard works with the help of the keyboard controller chip present on the circuit board of the keyboard. The keys of a keyboard can be of various types including membrane, dome-switch, scissor-switch, capacitive, mechanical, and so on. Among all these types of keys, the mechanical switches like Cherry MX are often used in most keyboards for desktop computers. The circuit board of the keyboard consists of a matrix of such mechanical switches arranged according to the QWERTY keyboard layout.

This matrix of keys are connected to the keyboard controller chip through a bus. Whenever a key is pressed, the keyboard controller would decode the key and send a signal to the computer as a scancode. The computer would then map the scancode to the relevant data, which could be a number, character, or any control command. The keyboard is connected to the computer through the host interface that could be either PS/2 or USB.

The onboard light indicators for "Num Lock," "Caps Lock," and "Scroll Lock" are controlled by the keyboard controller, and whenever any one of the lock keys are pressed, the corresponding indicator light is turned on.

Wireless keyboards that work with Wi-Fi and Bluetooth also include a keyboard controller but may also include more chips and complex circuit for radio communication and the inbuilt touchpad controller.

Specifications of Keyboard

There is nothing much about the specifications of a keyboard but it can include a few like the total number of keys, interface, and cable length.

- Total keys—this is the total number of keys present in the keyboard and a standard keyboard should have at least 104 keys.
- Interface—it's the connectivity standard through which a keyboard would connect with a computer. Earlier, it was PS/2, but now most keyboards can be connected through USB interface only.
- Cable length—in case of a wired keyboard the cable length matters a lot and the length would depend on the location of the desktop computer. Longer the cable, better is the flexibility, but too long could cause the cable to tangle more often. That's why a cable length between 1 and 2 meters is ideal for use with a desktop computer.

Mouse

A mouse is a pointing device used with a computer to interact with the GUI. With the help of a mouse one can operate the computer in a user-friendly manner, without the need of typing any commands through the keyboard. It was first invented by Douglas Engelbart during the 1960s and was later developed by Palo Alto Research Center (PARC) to use it with a computer called Xerox Alto in 1973. Since then many companies like Logitech have developed and manufactured various types of mouse for the consumer market (see figure 6.1).

Earlier, the mouse used to come with a rubber ball and the movement of the ball was tracked by the sensors of the mouse, which in return would tell the computer where to move the mouse pointer. Later this technology was replaced by the optical mouse, which uses a camera with light to detect the movements.

Components of an Optical Mouse

The optical mouse mainly consists of buttons, scroll wheel, camera, LED, and a controller chip. The LED light travels through a transparent block of plastic that causes internal reflection and lets the light travel to the surface below the mouse, and any mouse movements are detected by the controller chip, through the camera (placed just above the surface). Both left and right buttons are also connected with the controller. The scroll wheel has slots through which light can pass (see figure 6.2).

The wheel is placed between an infrared LED and a light sensor. When the wheel rotates, the light is passed and blocked one at a time, and each time the

Figure 6.1 The Xerox Alto Computer with a Mouse. https://commons.wikimedia.org/wiki/File:Xerox_Alto_mit_Rechner.JPG

wheel rotates the number of light detection is converted as scrolling movements. Both the left/right button clicks and scroll wheel movements, along with the movement of the mouse (detected by the camera), are sent to the computer as scancodes, which are then mapped by the computer to place the mouse pointer on the screen or activate any event like popup of any menu or scrolling of a page in the GUI environment.

A mouse can also be used to play video games, and there is also a special gaming mouse available that comes with extra buttons for game control.

Specifications of Optical Mouse

An optical mouse can have specifications similar to the keyboard, and it includes a number of buttons, interface, cable length, and resolution range.

- Number of buttons—it's the total number of buttons present in a mouse. A normal mouse contains at least three buttons—left, right, and middle buttons. But any other multimedia and gaming mouse can have more buttons

What Are Computer I/O Devices? 53

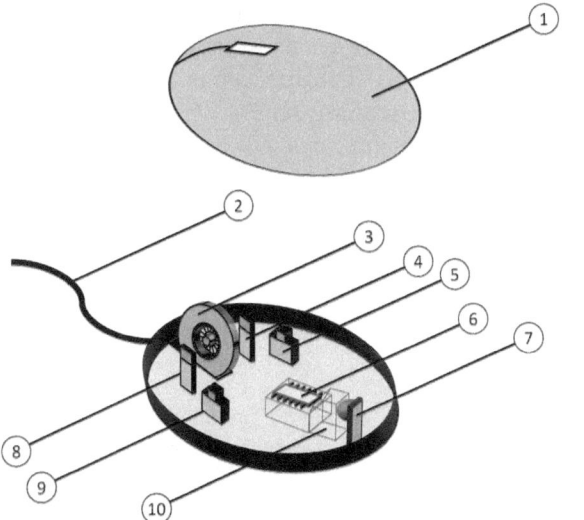

1) Top cover
2) Cable
3) Scroll wheel/middle button
4) Infrared LED
5) Right button
6) CMOS camera
7) LED
8) Infrared sensor
9) Left button
10) Lens with reflector

Figure 6.2 **Components of a Mouse.** Author Created.

for optional functionality. A mouse with three buttons for left, right, and middle clicks is sufficient for operating a computer.
- Interface—it's the connectivity standard through which a mouse would connect with a computer. Earlier, it was PS/2, but now most mice can be connected through USB interface only.
- Cable length—in case of a wired mouse, the cable length should be sufficient enough so that it can be moved freely. The smoothness and flexibility of the cable can also affect the usability—if the cable is too hard and sturdy, then it will become difficult to move the mouse freely, and can hamper any productive work that needs precise mouse movements. That's why a mouse with a smooth cable can be a better option, and a cable between 1 and 2 meters is ideal for use with a desktop computer.
- Resolution range—it's the minimum and maximum resolution of the optical mouse measured in dots per inch (DPI). Wider the range, better is its operating modes. A mouse with low resolution mode will work with less sensitivity and that may be required while playing video games and using GUI operating systems, but a higher resolution may be needed while doing productive work like photo editing. That's why a mouse with a resolution range of 800–1600 DPI is ideal for normal use, but 200–4000 DPI is best. However, most manufacturers don't mention the resolution with other specifications.

Joystick or gaming pad is also an input device that can be used with a computer, but it's mainly used for playing video games.

The two input devices—keyboard and mouse—have been used mainly to operate a computer. And much more research is underway to develop better input devices that can be used to control a computer more easily. A device called the Brain Computer Interface (BCI), which uses sensors to detect brain waves to control the computer, is still being researched for it'scommercial use. With the help of a BCI device one can control a computer directly with the mind. Such a BCI device needs to be in contact with the head of the user to pick up the brain waves. Because of that reason a BCI device is not convenient to use and hasn't got much commercial success yet.

OUTPUT DEVICES

There are many types of output devices that can be used with a computer, but the most commonly used one is an LCD/LED display and a printer. But it's not possible to operate a computer using a printer in real time, and that's why a display monitor is used. Earlier, CRT monitors were used as displays to operate the computer; now LCD/LED displays are mostly used because of their low power consumption and better picture quality.

LCD/LED Display

An LCD/LED display is an output device that can show images generated by a computer. With the help of an LCD/LED display it's possible to operate a computer in real time. An LCD display uses liquid-crystal technology to generate images on the display screen, whereas an LED display uses LCD with an LED backlit as the source of light.

LCD technology works by filtering unpolarized light using polarizing filters along with a material called liquid crystal. Any light emitting from a source like an incandescent lamp or LED can be unpolarized, which means that the light waves can vibrate in both vertical and horizontal directions. On the other hand, a polarized light can vibrate in either vertical or horizontal direction.

In an LCD, the liquid crystal is sandwiched between the two polarizing filters (both vertical and horizontal) that are aligned at right angles and when light passes from the first polarizing filter, only one type of polarized light is allowed to pass and the liquid crystals that is present between the polarizing filters, have a twisted structure, which also works like a polarizing filter. So, when the light passes through the twisted liquid crystal, it gets polarized again and as the polarization matches with the second polarizing filter, the light gets passed through it.

The twisted liquid crystal is placed between two glass substrates with transparent electrodes, and when an electric field is applied through these electrodes, the structure of the liquid crystal changes and becomes untwisted. This changes the polarization of the light passing through the liquid crystal, but doesn't match with the second filter and the screen appears black. In a LCD display images are created using pixels and each pixel is a cell containing electrodes with liquid crystal.

However, a color LCD contains additional RGB filters in every cell for red, green, and blue colors to control the color depth. As this type of LCD contains twisted liquid crystal, it's also called Twisted Nematic (TN) LCD.

The first LCD was invented at the Radio Corporation of America (RCA) Laboratories in 1964, but the TN LCD was invented by Martin Schadt in 1970. Later many Japanese watch manufacturers like Seiko used the TN LCDs in digital wristwatches during the 1970s (see figure 6.3).

There are two types of LCDs—passive and active. A passive LCD is the oldest technology in which each pixel on the screen is arranged in a matrix and a grid of vertical and horizontal conductors are connected with the pixels. The intersection of each vertical and horizontal conductor can control a pixel to either switch it ON or OFF.

Whereas, in an active-matrix LCD each pixel is controlled using a Thin-Film-Transistor (TFT) and the depth of the color can be precisely controlled with voltage. That's why such an LCD is also called a TFT display. Both passive and active LCDs can display color by making each pixel with three

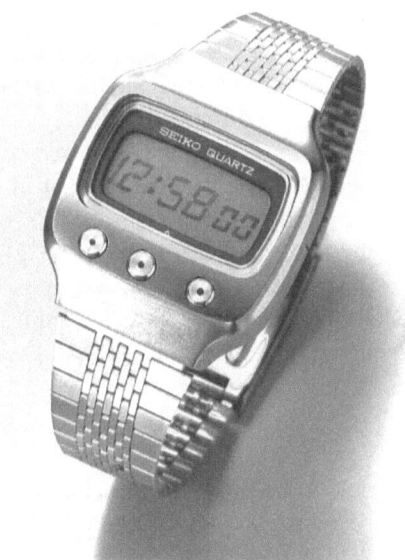

Figure 6.3 The World's First Six-Digit LCD-Based Wristwatch (06LC) from Seiko. Authorized by Seiko.

equal parts with color filters that would either block or pass only red, green, and blue (RGB) light all at the same time, but in different ratios to produce individual colors.

One problem with the TFT LCD is that it can't display good picture quality and has a narrow viewing angle. That's why another improved LCD called the In-Plane Switching (IPS) LCD was developed by Hitachi in 1992. In an IPS display the liquid-crystal rods are arranged in a way that the light passing through them can be seen from every angle, and that's why IPS displays have better viewing angles.

Earlier, most TFT LCDs used an incandescent lamp as a backlight to light up the LCD but later it was replaced with LEDs and that's why nowadays most LCDs with LED backlit are called LED displays, but they can still be TFT. Most IPS displays also have LED backlight but are better than TFT displays.

Another type of display called Active-Matrix Organic Light-Emitting Diode (AMOLED) was developed by Samsung in 2006 that used RGB LEDs for each pixel to display a color picture. The advantage of AMOLED display is that it doesn't need any backlight and each pixel can be controlled individually to display high-quality pictures with precise color gamut.

Touch screen displays can also come with variants like TFT LCD/LED, IPS, AMOLED, and so on, but it also works as an input device because it can take the touch responses as input data for the computer.

Almost all types of displays have a refresh rate that is measured in hertz (Hz). The video is generated by the computer as a frame and the GPU can create many frames in one second, which is called the frame rate. The frame rate is measured in frames per second (fps), and the frame rate of most desktop computers can vary from 24 fps to 60 fps. So, the display device also needs to display each and every generated frame in a timely manner, and needs to be synchronized with the computer. So, the display screen needs to be refreshed every time to display a new frame, and the number of times the display gets refreshed in one second is called the refresh rate.

Usually, the refresh rate of most displays can be 60 Hz, but there are displays with higher refresh rate also, and can go up to 240 Hz. Such higher refresh rates are required for playing HD video games, but the GPU should also be capable of generating the video at 240 fps.

Components of an LCD

An LCD mainly consists of polarizing filters, liquid-crystal, glass substrate/electrodes, color filter, and backlight (see figure 6.4).

- Polarizing filters—it's the filter that changes the polarization of light passing through the LCD. First the unpolarized light is polarized using the first

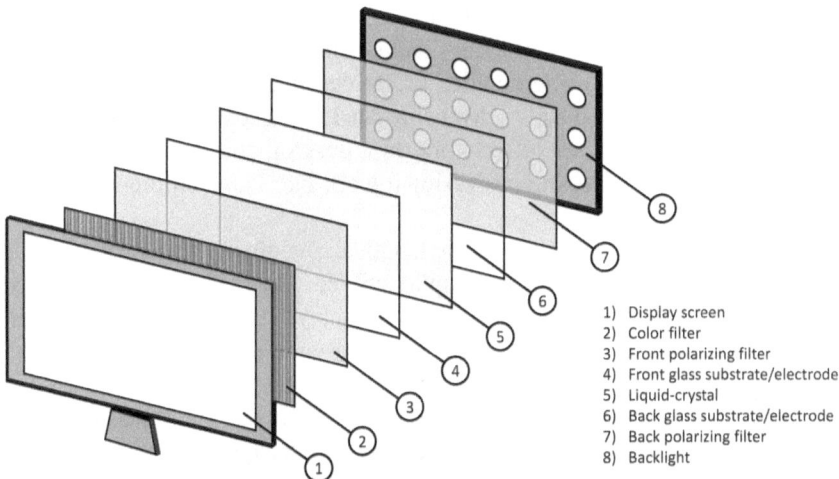

Figure 6.4 Components of an LCD. Author Created.

polarizing filter and then the second filter changes the polarization of the light again to make the LCD cell either black or color.
- Liquid-crystal—this is the main component of the LCD that makes changes to the light. It's the middle layer of the LCD and it remains in twisted condition. But becomes untwisted when an electric field is applied. It works like an electric filter that can either let the light to pass or block with different voltage levels.
- Glass substrate/electrodes—it's a glass with transparent electrodes made from Indium Tin Oxide (ITO) and the liquid crystal is placed between them. The front glass contains electrodes with the shape of pixels and when voltage is applied to the electrodes of the two glass substrates, the liquid crystal becomes untwisted and changes the polarization of the light.
- Color filter—it's a filter for red, green, and blue colors of the LCD. With the presence of this filter different colors become visible on the LCD screen.
- Backlight—it's a light source that is present on the back side of the LCD and without it LCD can't display anything. Usually, most TFT LCDs use incandescent lamps as a black light, but most modern LCDs use LED instead, which are also called LED displays.

Specifications of LCD

An LCD can have specifications like size, aspect ratio, resolution, refresh rate, interface, and power consumption. However, these specifications can also be applicable for LED, IPS, and AMOLED displays also.

- Size—it's the diagonal size of the LCD and it's measured in inches. Most desktop LCDs start from 20 inches, but smaller than 20 inches may be also available. However, using smaller LCDs can make things difficult while using the computer. That's why using an LCD between 21" and 27" is ideal for doing home or office work on a desktop computer. Bigger-sized LCDs are useful for doing professional work like video editing or playing HD video games.
- Aspect ratio—this is the ratio of the length and width of the display and this ratio can also tell how wide the display will be. Most HD videos are played with 16:9 aspect ratio and that's why a display with that aspect ratio will be a standard one. However, there are ultra-wide monitors available with higher aspect ratio like 21:9 for gaming and productive use.
- Resolution—it's the highest resolution the display can support, and higher the resolution, better is the picture quality. Nowadays, most videos can be played at FHD, and the LCD should have a resolution of at least 1920 × 1080, but higher resolutions like 3840 × 2160 can be used to support 4K UHD video. However, the HDMI has many versions and the latest one like 2.1 is better.
- Refresh rate—it's the rate at which video frames are displayed and it's measured in hertz (Hz). Higher the refresh rate, better is the video stability. Most displays support 60 Hz to 75 Hz, but LCDs with higher refresh rates are also available for gaming and other use.
- Interface—this is the connectivity standard through which the display can be connected with a computer to receive the video. Earlier, the interface was VGA but now it's HDMI. However, some displays can also have a VGA interface for backward compatibility.
- Power consumption—it's the maximum power consumption the LCD can consume while operating in normal condition, and it's measured in watts (W). Lower the watts, more energy efficient is the LCD. Usually any standard LCD monitor for a desktop computer can consume power between 35 and 40 W.

There are many more output devices that can work like a computer display and one such device is a digital projector, which can show the video display projected on a white screen, just like a movie projector. Such a projector is mostly used in offices for presentation purposes during business meetings. Apart from the projector, there is another display device called the virtual reality (VR) headset that can be used to experience 3D video generated by the computer. VR headsets are mostly used for gaming and virtual training purposes.

Chapter 7

What's an Operating System (OS)?

The operating system (OS) of a computer is a software that is responsible for controlling the hardware and running the applications for the user. Without the OS, a computer cannot work independently because the instructions to the computer hardware are given by the OS, and various hardware components like the CPU need instructions from the OS in order to do all kinds of processing.

The OS also makes it possible to use multiple applications on a computer, which was not possible on previous computers like early mainframes that didn't have any multitasking OS. Most OS for general-purpose computers come with a kernel, which is the core of the OS that controls the hardware and helps the applications to use various resources of the computer. An OS, like Windows, comes with device drivers that let the OS control a specific hardware, and it's like a system software that gives instructions to the hardware.

Earlier, the device drivers for most OS, like Windows, needed to be installed manually, but after the introduction of the Plug and Play (PnP) feature, the drivers can now automatically get installed by the OS.

Most OS of general-purpose computers come with user interfaces like the CLI or GUI. Earlier OS like Unix and Linux came with CLI, whereas modern OS like macOS and Windows have a GUI that makes them user-friendly and commercially successful.

OS are developed using software engineering tools and programming languages (used for coding) like C and assembly language. The source code of many OS like Unix and Linux are under public license and can be modified for specific requirements. But the source code that is under Berkeley Software Distribution (BSD) license can also be used for commercial development, and that's how Apple used most of the code from two Unix OS distributions

(NeXTSTEP and FreeBSD—under BSD license) to develop the macOS. Even Sony used the FreeBSD OS code to develop the PlayStation OS.

Earlier mainframe computers didn't have any OS but only required a single program to run from a punched card or magnetic tape. The first OS was the GM-NAA I/O for IBM 704 mainframe computer, created by Robert L. Patrick of General Motors Research and Owen Mock of North American Aviation in 1956.

Later, IBM developed and released many OS like OS/360 and Disk Operating System (DOS)/360 for the OS/360 system. In 1962, Eckert–Mauchly Computer Corporation (manufacturer of the first commercial computer and now known as Unisys) developed the EXEC I OS for the Universal Automatic Computer (UIVAC) 1107.

After the arrival of minicomputers, many companies developed OS for their minicomputers. Companies like Digital Equipment Corporation created OS like RT-11, RSX-11 for its 16-bit PDP-11 minicomputer, and VMS (now available as OpenVMS from VSI) for 32-bit VAX machines.

IBM also created the OS/400 for the Application System(AS)/400 minicomputers.

The Unix operating system was developed at AT&T Bell Laboratories in the late 1960s, originally for the PDP-7, and later for the PDP-11, which were also minicomputers.

The use of computers changed after the introduction of microcomputers in the market during the mid-1970s. Personal computers like Commodore64 and Apple II came with an OS that was able to control the hardware like a modern OS.

However, the OS called KERNAL that came with Commodore64 worked at low level like the basic input/output system (BIOS) of a desktop computer, whereas Apple II was able to run the OS called AppleDOS, which could run other applications like Wordstar and dbase II.

Apple II was also able to run third-party OS like the Control Program/Monitor (CP/M) and Graphic Environment Operating System (GEOS). Based on the CP/M, Tim Paterson developed the 86 DOS (also known as QDOS), and Microsoft hired Tim in 1981 and bought the 86DOS for US$ 75,000.

Microsoft renamed it to MS-DOS and also licensed it to IBM as PC-DOS for their computers. It was 86DOS of Tim that helped Bill Gates to grow Microsoft's business globally.

Both MS-DOS and PC-DOS have CLI as the user interface, and the user needs to enter DOS commands to operate the computer. Earlier, computers used to start using MS-DOS, and Windows OS was executed to run from the DOS shell, but now computers start with a Windows OS and MS-DOS can be executed to run from the GUI.

The GUI-based OS was first developed by Xerox at the PARC for the Xerox Alto computer 1973. Later, many companies developed similar GUI

OS for their computers, such as Apple's macOS, the Radio Shack Color Computer's OS-9 Level II/MultiVue, Commodore's AmigaOS, Atari's TOS, IBM's OS/2, and Microsoft's Windows. Nowadays most OS that run on many other devices like smartphones also have a GUI and come with user-friendly applications.

The Android OS from Google has been developed from the Linux kernel, and there are many other newer OS that were developed from the Linux and Unix kernel as well.

Just like any CPU can have a bit-length (32-bit or 64-bit), OS for most x86-64 computers can be either 32-bit or 64-bit. A 64-bit OS can let the application take full advantage of a 64-bit CPU, and only those applications that are designed to work with a 64-bit CPU can only benefit from a 64-bit OS but not all applications. A 32-bit OS can run on both x86 and x64 systems, but a 64-bit OS can only run on a x64 system.

TYPES OF OS

OS can be of various types, and the most common types include single-tasking and multitasking, single and multiuser, network, distributed, embedded, and real-time.

- Single-tasking and multitasking—a single-tasking OS can run only one program at a time, whereas a multitasking OS can run more than one program at a time. Multitasking OS works using time slicing method, in which every program is run at the same time by repeatedly interrupting and scheduling the runtime based on priority. MS-DOS is an example of a single-tasking OS, whereas Windows and macOS are multitasking OS.
- Single and multiuser—a single-user OS can support multitasking but cannot allocate processes and resources to multiple users individually at the same time, whereas a multiuser OS can. A multiuser OS can identify each process and resources like devices and storage space belonging to a particular user, and can also let multiple users access the system at the same time. DOS is an example of a single-user OS, whereas Unix is a multiuser OS.
- Network OS—a network OS is designed to work with servers and networking devices like routers and switches. Such an OS comes with many security and networking features required for running a computer on a network. Windows Server 2022 is an example of a network OS used in many servers, and Cisco Internetwork OS (IOS) is used on Cisco Systems routers and network switches.
- Distributed—it's an OS that combines more than one computer on a network and makes them appear as a single computer, while keeping the work

distributed among the connected computers. MOSIX and Sprite are some examples of distributed OS.
- Embedded—the OS that is used on embedded computer systems is called embedded OS. They are designed to work efficiently on computers with limited resources. Windows CE and Minix 3 are some examples of embedded operating systems.
- Real-time—this type of OS has better performance and guarantees to process any event or data in a specific time. It can be either a single or multitasking but performance is maintained by using specialized scheduling algorithms.

OS ARCHITECTURE

Like any computer hardware architecture, any software can also have an architecture that includes the OS too. However, the architecture of any OS depends on the design of the kernel. Depending on the design, an OS can have monolithic, micro, and hybrid kernels (see figure 7.1).

- Monolithic kernel—OS with this type of kernel runs the services and programs within the kernel's memory space. The kernel works like a hardware interface for the applications and this design provides rich and powerful hardware access. One big disadvantage of this design is dependencies of the system components, and if any component fails then the whole system crashes. AIX, HP-UX, and Solaris are some examples of OS with monolithic kernels.

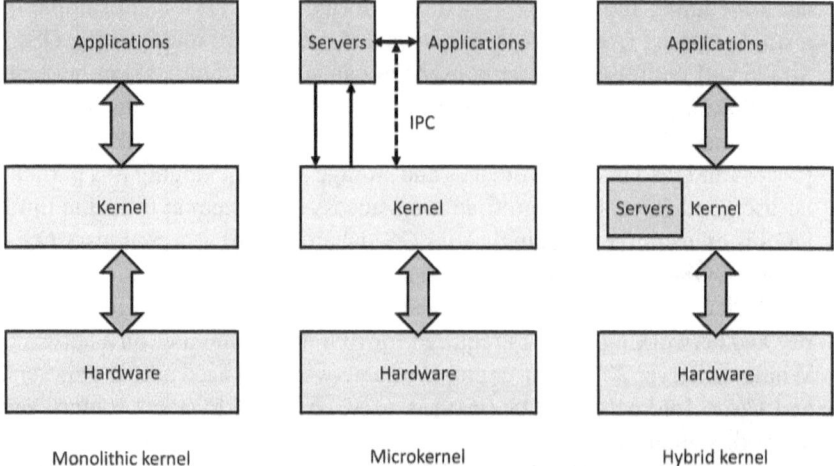

Figure 7.1 Three Different Architectures of Kernel. Author Created.

- Microkernel—in this design, the kernel runs in a separate memory space providing basic functionality only. The other advanced functionalities like networking are implemented in user-space programs referred to as servers. Applications running in the user space send requests to these servers as Inter-Process Communication (IPC) and a server fulfills the request (like hardware access) through the kernel.

  This kernel architecture makes the system more stable and makes it free from crashes because the kernel runs independently within a separate space from the user applications. The one big disadvantage of this design is that a large number of system calls and context switches can slow down the system because that can generate more overhead than plain function calls. It also lets the kernel take less space—making the kernel become smaller in size, and that's why it's called microkernel. HelenOS, Horizon, and Zircon are some examples of microkernels.
- Hybrid kernel—this kernel design combines both the monolithic and microkernel architecture. In this design, some important services like network stack and filesystem run within the kernel space (like the monolithic kernel). But some other kernel code, like device drivers, run as servers in the user space (like the microkernel). This reduces the possibility of crashes from unreliable services and also minimizes the overhead from large system calls and context switches. Most modern OS are designed with hybrid kernel, and Windows, macOS, and FreeBSD are some examples of OS with hybrid kernel.

SYSTEM REQUIREMENTS

Any software requires hardware to run, and any OS also requires a computer to have minimum hardware requirements to run properly. Hardware requirements include processor, memory (RAM), storage space, firmware, and GPU.

- Processor—a processor can have hardware architecture, bit-length, and frequency that should be supported by the OS. Computers can have various architectures like IA-64, x86-64, and Arm64. An OS that is designed for a computer with a particular hardware architecture can run on that hardware only. For example, Windows OS made for x86-64 computers like desktops and laptops can't be installed on smartphones with Arm64 architecture, but could run on some tablets with x86-64 architecture.

  Processors also have bit-length—it's the bit-length of the CPU and usually most OS for x86-64 CPU comes in either 32-bit or 64-bit versions. A 64-bit OS can run on a 64-bit CPU but can't run on a 32-bit CPU. However, a 32-bit OS can run on both 32-bit and 64-bit computers. Nowadays, most modern CPUs used in desktop computers are 64-bit wide and support a 64-bit OS like Windows 11.

A CPU should have the required minimum operating frequency for the OS to run properly and the number of cores also matters. Most latest versions of general-purpose OS like Windows 11 require at least 1 GHz of CPU frequency with two cores.

- Memory (RAM)—it's the amount of RAM (measured in bytes) required for the OS to run properly. Higher the memory, better is the performance of the OS. Nowadays, most OS like Windows 11 require 4 GB of RAM to run.
- Storage space—it's the amount of space (measured in bytes) required in the secondary storage device like the hard drive (where the OS needs to be installed). Unnecessarily increasing the storage space above the required amount won't improve the system performance. Most operating systems like Windows 11 require 64 GB of storage space.
- Firmware—it's the software that remains in the ROM chip of most computers, and is required to give initial instructions to the computer on how to boot the OS. Booting is a process through which the kernel is loaded into the computer memory and makes a computer fully functional. Without the firmware a computer won't be able to boot the OS.

  A desktop computer comes with two types of firmware, legacy BIOS and Unified Extensible Firmware Interface (UEFI), whereas UEFI is the latest type of firmware available in most recent computers. Windows 11 supports only UEFI and that's why it can't be installed on computers with legacy BIOS.
- GPU—it's the type of graphics processor present in the computer and the OS that supports a particular type of GPU should be present in the computer to run it properly. Most Intel processors come with integrated GPUs and are supported by most modern OS with GUI like Windows and Ubuntu. However, a GPU that supports the latest versions of OpenGL and Direct X can also support the latest OS with GUI.

The system requirements for an OS like Windows is required because it can be installed on any custom-built computer hardware. However, this is not the case for macOS because it always comes with an Apple computer that has matching hardware. Before installing any OS the storage drive should be formatted with a suitable file system like FAT32 or NTFS for Windows, GPT for Linux, and APFS for macOS.

Chapter 8

What's Application Software?

A computer application software or app is one or more programs that makes a computer perform a specific task—defined by the user. A general-purpose computer can perform various types of tasks based on the type of apps used. It's the app that can make a computer switch its functionalities, and a user can make use of a computer by choosing the app as per the requirements. For example, a computer can be made to process the scores of students in a school by choosing a spreadsheet app like Microsoft Excel or the computer can be made to work as a video calling device by using an app like Skype.

Apps are created using a Software Development Kit (SDK) like Android NDK, iOS SDK, Java Development Kit, and Microsoft Windows SDK, and an Integrated Development Environment (IDE) like Microsoft Visual Studio or Apple's Xcode that supports programming language like C/C++, Java, Python, .Net, and Objective-C. SDKs are like a package that comes with all the software tools required for building an app, whereas IDEs are like interfaces for the programmer to code and test the app.

The code (in any programming language) used for building an app is converted to a machine-readable code called as a bytecode or executable code, because a code written in any programming language is either mid-level or high-level language and contains expressions that are meaningful to a programmer only. A computer can only understand binary code (0's and 1's). That's why the programming code must be converted to a binary code. This is done with the help of a compiler that comes with the IDE. In Windows, a compiled program gets saved with an EXE file extension, but this is not the same for other OS like macOS or Android.

An app can be created with the same algorithm in many different programming languages, but after compilation it will give the same results. It depends on the programmer in which language the app should be coded.

Algorithms are expressions written in the English language—to show the logic and control flow of a program—before coding it with a programming language.

A computer program can be written in many different languages but with the same algorithm and logic. For example, the sum of two numbers can be calculated using three different programming languages as follows:

- Using BASIC:
  Dim x as Integer = 1
  DIM y as Integer = 2
  DIM z as Integer
  z = x + y
  print "Sum of x + y ="; z

- Using C:
  #include <stdio.h>
  int main()
  {
  int x, y, z;
  x = 1;
  y = 2;
  z = x + y;
  print ("Sum of x + y =%d," z);
  }

- Using Python:
  x = 1
  y = 2
  z = x + y
  print("Sum of x + y =", z)

Apps can be either created as a installation package using a software framework or can be a simple program written using a procedural language like Beginners' All-purpose Symbolic Instruction Code (BASIC) or C, whereas an app with multiple programs can be created using Object-Oriented Programming (OOP) languages like Java or Python. With procedural programming language, a program can be created with procedures—containing computational instructions to be carried out in steps, but with OOP language, an app can be created with multiple procedures by creating objects.

Both procedural and OOP languages fall under imperative programming language, which tells the computer how to do things with logical expressions and defines the structure of the program. On the other hand, the declarative type of programming languages doesn't tell exactly how to do a particular task but only expresses the logic without defining the structure of the program. Structured Query Language (SQL) is a declarative language used for programming and managing Relational Database Systems (RDMS) like MySQL or Oracle database systems.

The imperative language is like calling a friend and telling him how to do math homework, but a declarative language is like calling a friend (who is already an expert in math) to do the math homework without teaching him how to do it.

However, most applications are built using imperative programming languages like OOP because OOP languages provide much more flexibility. The framework and API are the two important abstractions that also facilitate many additional functionalities of a software development. A framework is like a foundation software that comes with compilers, libraries, toolsets, and APIs, which can be used during a software development process to change the basic functionalities provided by the framework—with the help of a programming language.

API makes it possible to build apps with additional functionalities or share functionality from other apps or services. It's an interface that lets a programmer communicate with other apps or services. With the help of an API an app can access the services and resources of another app without knowing about its internal functionality and code.

That's why while developing an app it's important to choose the right SDK and programming language. The framework also plays an important role for the app creation because it provides the overall infrastructure for building an app. The Windows .Net framework makes it possible to compile a code written in many different programming languages like C#, J#, and Visual Basic. NET (VB.NET).

This is possible because the compilers for each language compiles the code to a Common Intermediate Language (CIL) and CIL is used by a virtual machine called the Common Language Runtime (CLR), which further compiles to machine code. Both the CIL and CLR together create an infrastructure called the Common Language Infrastructure (CLI).

Code written in some languages can be compiled to machine language using a compiler, but in case of some programming languages, the code can run using an interpreter instead, which executes each line of a code at a time. Programming languages like BASIC and Python use interpreter.

Apps are designed by software engineers and coded by a programmer. The design is done for the architecture and data flow logic of the app. Most of the design is done using various Data-Flow Diagrams (DFD) and flow charts. The apps work with two things—data that is processed by the app, and algorithm that follows the logic. DFD is used to represent the flow of data and flowcharts are used to show the logic of the algorithm that the app will follow. Algorithms are just logical representations of instructions that will be used while coding a program or developing an app.

Apps for a general-purpose computer require some kind of operating system (OS) to run, and an app built for a particular OS can't run on another OS. For example, the Mozilla Firefox web browser made for Windows can't run on macOS, and the one made for macOS won't run on Windows. This is because apps have system dependencies, and the app that is compiled on a particular OS will run on that OS only. Apps run at the topmost layer between the user and the OS, and only with the help of the OS, apps can make use of other services.

The world's first computer program was written by Tom Kilburn and it ran on the Manchester Baby (world's first electronic stored-program computer) at the University of Manchester on June 21, 1948, and calculated the highest factor of the integer $2^{18} = 262{,}144$. The program was loaded to the computer using a punched card.

APP'S LIFE CYCLE

Any app goes through various phases of development—starting from analysis and design to the release of the app. When the app development process goes through all these phases and the app is finally released, it's said to have completed one life cycle. But the cycle may continue again and again for future releases (with updated versions) of the app with many improvements. A basic life cycle of an app, also called the Software Development Life Cycle (SDLC), consists mainly of five phases—analysis/requirements, design, coding, testing, and release/maintenance (see figure 8.1).

- Analysis/requirements—in this phase, analysis about the app is done and the requirements for the software development like which SDK and programming language needs to be used are decided. Also the duration of the development project is estimated and everything is scheduled as per the timeline.
- Design—after everything is planned, the design phase starts, and here the logic and the architecture of the app is understood using various tools, and various documents like flowcharts and DFDs are created.
- Coding—in this phase, the actual code is written and a working prototype of the app is created.
- Testing—after the coding phase is over, the prototype of the app is tested in this phase. Testing is done using various testing tools and methods like black-box and white-box testing.

- Release/maintenance—after when the app is fully tested and works properly, it is released with a version number. But if any bug (a glitch in the code) is reported by the user, then either a patch is released to counteract the bug or the app goes through another iteration of all the phases, and another version of the app is released with a new version number. In this manner the app is improved in every cycle either to remove any bug or to add new features.

TYPES OF APPS

Based on the usage and requirements, apps can be classified as information worker app, entertainment app, communication app, enterprise app, simulation app, media development app, product engineering app, software development app, and educational app.

- Information worker app—this type of app can process information like documents and databases. They are mainly used for processing and storing information for business and educational purposes. Microsoft Office and MySQL are some examples of information processing apps.
- Entertainment app—with this type of app the user can get entertainment like playing video games, watching movies, and listening to music. Video games are a type of entertainment app that requires the power of a graphics processing unit (GPU) to play and sometimes requires additional input devices like a gaming pad for playing. Minecraft and Tetris are examples of some popular video games.
- Communication app—this type of an app is used for communication between computers and also for sharing or accessing data over a network or the internet. For example, a Voice Over IP (VoIP) app can be used to make voice calls over the internet, and a web browser app can be used to access information on a web page from a computer (server) connected to the internet. Skype, WhatsApp, Messenger (Meta), Telegram, ICQ, and so on, are some examples of VoIP apps. Google Chrome, Apple Safari, and so on, are some examples of web browsers.

    Apps that can play a video are called media players and they are used for playing online or offline videos; they are also used for watching movies. VideoLAN Client (VLC) player and Media Player Classic—Home Cinema (MPC-HC)—are some examples of video players. Apps can also be used for playing audio and music for entertainment.

    There are two types of audio players—one for playing audio files and other for Musical Instrument Digital Interface (MIDI) files. Any audio file

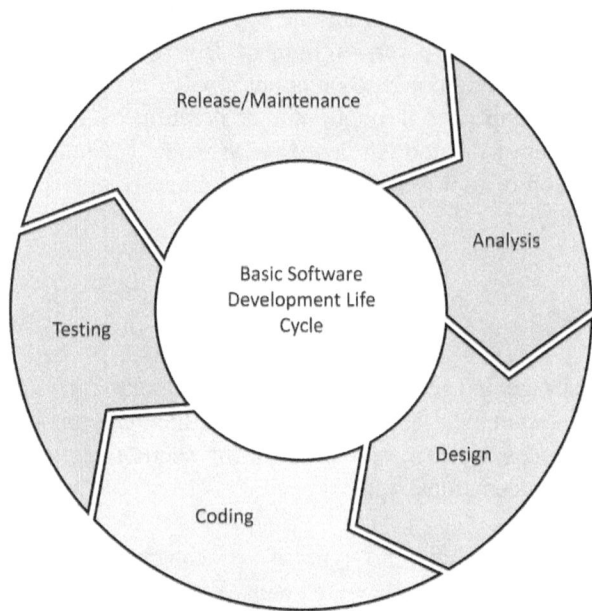

Figure 8.1 A Basic Software Development Life Cycle (SDLC). Author Created.

can be of various formats (WAV, FLAC, MP3, etc.) and can contain any kind of audio, but a MIDI file can only contain music.

Apps like AIMP and iTunes are some examples of audio players. VanBasco's Karaoke Player, Soundfont Midi Player, and Yamaha S-YXG50 VSTi Synthesizers are some examples of MIDI players.

Image file viewers can also be used for entertainment to view images as slide shows, and most of them come with the OS.

- Simulation app—with this type of app any physical system of the real-world can be virtually created to predict the behavior of the system under certain conditions and can be used to interact with the virtual system for training purposes. For example, a simulation app can be used to predict and solve multiphysics problems used in engineering but a flight simulation app can train someone how to fly an aircraft.

ADINA is a simulation software used in engineering to solve structural, fluid, heat transfer, and electromagnetic problems, and Microsoft Flight Simulator can be used to simulate an aircraft on a computer. Such flight simulators and other types of game like simulators also fall under entertainment apps.

- Media development app—this type of app is used for productivity and content creation. Such apps are used for video/audio editing, music production, photo editing, and graphics. Apps like Adobe premiere and

Magix Vegas Pro are used for video editing, and apps like Magix Sound Forge, Audacity are used for audio editing. Apps like FruityLoops Studio, Steinberg Nuendo, and Magix music maker are used for music production purposes. Photo editing can be done using apps like Adobe Photoshop and CorelDRAW. Vector processing apps like Adobe Illustrator can be used for graphics design.

- Product engineering app—with this type of app engineers can design products for commercial and research purposes. Computer aided design (CAD) apps like Autodesk Fusion 360 are used in engineering for designing products.
- Software development app—this type of app is used for creating any kind of app or system software using any programming language. Such apps include IDE, code editors, and compilers. CPython and Microsoft Visual Studio are examples of software development apps.
- Educational app—this type of app is used for educational purposes and is used for teaching and learning purposes. Such apps are available based on the educational subjects and can be used by teachers and students related to that subject. For example, an app for solving mathematical problems can be used by teachers and students dealing with that subject, and another app that can simulate physics, can be used for learning physics. Microsoft Math Solver (used for learning math) and Algodoo (used for learning physics) are some examples of educational apps.

Apps can provide the interfaces that the user can use to interact with the app, and every app should have a user interface (UI). An app with a good architecture and UI can let the user use a computer effectively. Therefore, the quality of an app can affect the usability and performance of a computer, and choosing the right app can be challenging sometimes.

Chapter 9

What's a Computer Network?

A computer network is a group of computers connected with each other in a way that they can exchange data with each other. Any computer or device connected to a network is called a node and each node is identified with a physical address called the MAC address and an IP address. The MAC address is used to identify the physical presence of a node on a network, and it consists of 48-bit six group of hexadecimal numbers separated by colons, hyphens, or without any separator (e.g., 00:00:5e:00:53:af).

Usually any form of storage that holds data is called a media, and on a network, computers can be considered as a media because they can store data. That's why accessing a computer's data on a network is called media access, and the process of reading and writing data to such media needs some kind of control. Controlling the media access over a network is called media access control (MAC), which is done by the Network Interfaces Controller (NIC), like an Ethernet network card.

But to access the media of any node with a NIC on a network, it must be first located. That's why a MAC address is required, and without that media access is not possible. A MAC address is required for every device connected to a network, whether it's a wired or wireless network.

On the other hand, an IP address of a node is used to identify the network interface and find its location over a particular network. It has two versions—IPv4 and IPv6. IPv4 IP address is used for nodes in an IPv4 network, and the address consists of 32-bit value with four 8-bit decimal numbers separated by periods (e.g., 192.168.1.1). Each number can be between 0 to 255 and is also called an octet. IPv6 in a successor to IPv4 used for nodes in an IPv6 network, and the address consists of 128-bit value with eight 16-bit hexadecimal numbers separated by colons (e.g., 2a03:2880:f11c:8183:face:b00c::25de).

The first three fields (48 bits) represent routing or site prefix (public topology of a website), the fourth (16-bit) number represents the subnet ID (private topology of a website) and the last four fields (64 bits) represent the interface ID (derived from the MAC address using the modified EUI-64 format). For example, in the IPv4 address 2001:0db8:3c4d:0015:0000:8a2e:0370:7334, the first three numbers (2001:0db8:3c4d) represent routing prefix, the fourth number (0015) is the subnet ID, and the last four numbers (0000:8a2e:0370:7334) represent interface ID.

Network communication can be either unicast or multicast. When data is exchanged between only two nodes in a network, it's called unicast, and when one node sends data to multiple nodes, it's called multicast.

Depending on the cast, and network architecture, IPv4 addresses are grouped into five classes—class A, B, C, D, and E. Class A (addresses starting from 0.0.0.0 to 127.255.255.255), Class B (addresses starting from 128.0.0.0 to 191.255.255.255), and Class C (addresses starting from 192.0.0.0 to 223.255.255.255) are for unicast transmission.

However, Class D (addresses starting from 224.0.0.0 to 239.255.255.255) is for multicast transmission, and Class E (addresses starting from 240.0.0.0 to 255.255.255.255) is reserved and used for experimental and research purposes. IPv6 does not follow the classes of IPv4 and are grouped based on the primary addressing and routing methodologies common in networking—unicast, multicast, and anycast addressing.

MAC address is a physical address of the hardware interface (assigned at the time of manufacture by the manufacturer) and it helps identify a node physically on a local network, but an IP address is like a logical address of the node, and it helps to locate a node over one or more networks. A computer network can be either wired or wireless, and exchanges data in the form of packets.

Each packet contains two types of information—control information and data (payload). The control information consists of three headers—Ethernet header (containing the source and destination MAC address), IP header (containing IP address of source and destination), and TCP header (containing the port numbers of the source and destination along with other transmission control information).

When any data is sent from one computer to another through a network, it's first converted into network packets. For instance, if any file stored on a computer needs to be transferred to another computer through a connected network, it is first sliced down into chunks of data and each segment of data is packed with control information (headers), which is sent with a unique number (port number) within the TCP header that identifies the endpoint connection (port). This packaging of data with such control information creates a packet, which is sent from node to node to exchange data..

The network devices like routers and switches of the network reads the control information and forwards the packet to the destination node and after receiving the packet, the data segment is stored and the computer waits for the next packet. In the same way all the remaining chunks of data are sent as packets to the dedicated port number. After receiving all the data segments, they are converted back to a file. This method of sending data from one computer to another using packets is called packet switching, and it was first developed by Paul Baran in the 1960s.

There are various types of computer networks depending on the range and features. For example, nanonetwork and Near-Field Communication (NFC) are used to connect devices within a few millimeters, whereas a Metropolitan Area Network (MAN) and Wide Area Network (WAN) can have a communication range of several miles. On the other hand, a Virtual Private Network (VPN) and Storage Area Network (SAN) differ as per their features—VPN is used as a virtual network to connect with a private network through a public network using encryption methods, but SAN is used to access storage devices only.

Among all the types of computer network, the Local Area Network (LAN) is commonly used in homes and offices to connect computers within the same house or building. It can be either a wired (Ethernet) or wireless (Wi-Fi) connection, and can be only between Peer-To-Peer (P2P) using only two computers or a workgroup of many computers. Both the Ethernet and Wi-Fi network used in LAN follow the Institute of Electrical and Electronics Engineers (IEEE) 802.3 and IEEE 802.11 standards, respectively.

Most Ethernet networks use baseband for communication and a baseband uses a single channel to transmit or receive digital signal one at a time. Ethernet connection transfer rate is usually from 2.94 Mbit/sec to 400 Gbit/sec depending on the IEEE 802.3 version of Ethernet, and 1.6 Tbit/sec of upcoming versions. On the other hand, Wi-Fi (version 6 of 802.11ax) can have transfer rate between 600 and 9608 Mbit/sec on frequency bands of 2.4/5/6 GHz. The next Wi-Fi 7 (802.11be) will have a transfer rate of 40 Gbit/s on the same bands.

The transmission of data along with packets are done using some standard communications protocols known as internet protocols. There are many types of internet protocols used in network communication and are collectively called the internet protocol suite. This suite includes protocols like File Transfer Protocol (FTP), Hypertext Transfer Protocol (HTTP), TCP/IP, Internet Protocol (IP), User Datagram Protocol (UDP), and Datagram Congestion Control Protocol (DCCP).

But the foundational protocols of this suite include TCP/IP, IP, and UDP—TP/IP is used for transmission control, IP is used for routing packets over the internet, and UDP is used for data transmission of lossy data like audio/video

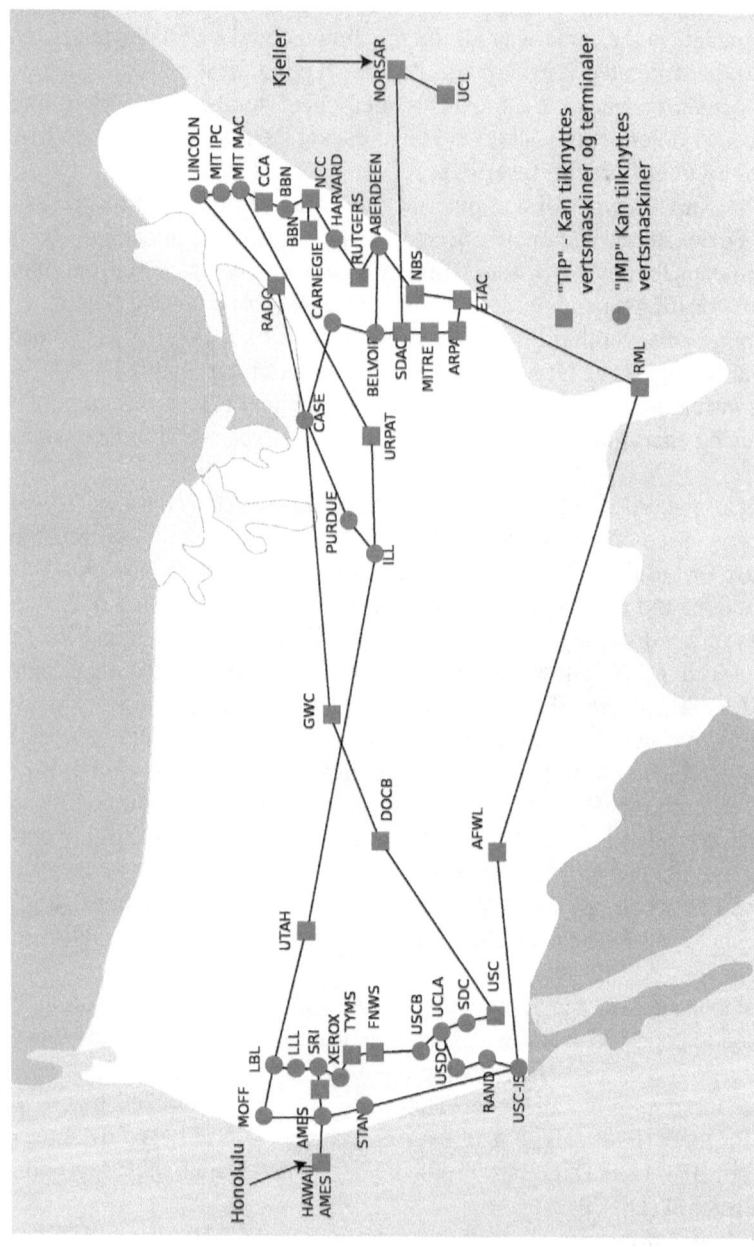

Figure 9.1 ARPANET Access Points in the United States in the 1970s. https://commons.wikimedia.org/wiki/File:Arpanet_1974.svg

streaming and for applications like voice over IP (VoIP) and online gaming. Computers in a network need a communication device like a router or a modem to communicate with each other.

In 1958 the American Telephone and Telegraph (AT&T) company released the first commercial modem called Bell 101 modem for the Semi-Automatic Ground Environment (SAGE) to connect a network of computers built for the US military. However, the Advanced Research Projects Agency Network (ARPANET) was the first WAN with TCP/IP protocol that was fully operational. The ARPANET project was led by Robert Taylor (computer scientist) in 1966, and after when the network was improved, it expanded rapidly and operational control passed to the Defense Communications Agency in 1975 (see figure 9.1).

THE OSI MODEL

Data from one computer to another in a network flows from the applications to the hardware of one computer (sender), and from the hardware back to the application of another computer (receiver). But from the application to the hardware, the data flows through seven layers, which are collectively called the Open Systems Interconnection (OSI) model. The OSI model is designed by The International Organization for Standardization (ISO). A network connection on a computer is first established by an application and the data is processed as per the required protocol.

- It then flows through many layers until it reaches the hardware layer, where it is physically transmitted over a network. The data actually experiences seven types of activities (while flowing through the seven layers) that follow various protocols defined in the internet protocol suite. The protocol for the data in each layer is selected based on the type of transmission selected at the application layer. The seven layers of OSI model include application layer, presentation layer, session layer, transport layer, network layer, data link layer, and physical layer.
- Application layer (layer 7)—it's the topmost layer of the OSI model, and here the applications give the user access to a network. In this layer the protocols like HTTP are used for high-level resource sharing or remote file access. HTTP, HTTPS, FTP, and SMTP are some examples of internet protocols used in this layer.
- Presentation layer (layer 6)—this layer converts the data based on the protocol that can carry on activities like data compression, character encoding/decoding, encryption/decryption. It works like a translator between the application and the network services. MIME, SSL/TLS, and XDR are some examples of internet protocols used in this layer.

- Session layer (layer 5)—in this layer the controls of the connections and terminations of a network transmission are managed. Here sockets are managed for the data transmission.
- Transport layer (layer 4)—this layer manages the transfer of data by means of various activities like segmentation, acknowledgment, and multiplexing. TCP, UDP, SCTP, and DCCP are some examples of internet protocols used in this layer.
- Network layer (layer 3)—it's a layer that manages the addressing of data and how it will be routed and delivered. IP, IPsec, ICMP, IGMP, OSPF, and RIP are some examples of internet protocols used in this layer.
- Data link layer (layer 2)—this layer is responsible for the transmission of data frames between two nodes connected by the physical layer and creates data link. PPP, SBTV, and SLIP are some examples of internet protocols used in this layer.
- Physical layer (layer 1)—it's the bottommost layer of the OSI model, and here transmission and reception of raw bit stream take place. It also defines the characteristics of the network hardware.

The four layers—application layer, presentation layer, session layer, and transport layer—are called host layers, and the three layers—network layer, data link layer, and physical layer—are called media layers.

NETWORKING DEVICES

A network with computers requires networking devices like hub/repeater, bridge, switch, router, and firewall to create the basic infrastructure of a network. That's why such a networking device connected to a network is called an infrastructure node.

- Hub/repeater—it's a networking device used to connect too many nodes (Ethernet devices) to create a network. A hub can have multiple RJ45 ports or BNC connectors that can be used with 10BASE-T or 10BASE2 networks. Hubs can also improve the network signals and can be used to extend a network, and that's why hubs are also called repeaters for Ethernet networks. Hub uses collision detection and fault isolation in the network, which improves the reliability of a network.
- However, there are other types of repeaters used in some networks of telecommunication, but for an Ethernet LAN, a hub can also work as a repeater.
- Bridge—this is a network device that connects two networks either wired (Ethernet) or wireless (Wi-Fi). A bridge for two Ethernet LAN networks can have two RJ45 connectors for connecting two 10BASE-T networks.

On the other hand, a wireless bridge can create a wireless connection with another wireless bridge or router using an ad hoc connection to connect two different networks (either LAN or WLAN).
- Switch—it's a networking device that works similar to a hub (i.e., to connect nodes) but it has better functionality than a hub. A hub works at the physical layer only and it sends packets to all the ports, and every node connected to a hub can receive the packets, which could take more bandwidth unnecessarily. Moreover, a hub is a half-duplex communication device, which means that only one node at a time can either send or receive data.
- Whereas a switch works at the data link layer and can learn about the MAC addresses of the packets, and it can send the packets only to that node for which the packets are addressed. This improves the network bandwidth and makes the network more secure. Apart from that some switches are full-duplex devices and have better routing functionality because they can even work on the network layer (layer 3) of the OSI model.
- Router—this is a network device that connects nodes to form a network and also helps in forwarding the packets in a network through the best available path, so that it reaches the destination node with least possible time. A router uses either wired (Ethernet) or wireless (Wi-Fi) connection to connect other nodes or networks. Usually a router is used to connect two or more switches that are connected to many nodes (forming a network).
- Firewall—this can be either a network device or a software that is used to filter incoming or outgoing connections of a network using network security and access rules. Usually a firewall is kept between a private and a public network to monitor incoming requests from a public network and can reject unwanted access requests from unrecognized sources to avoid any hacking attacks.

A modem is also a networking device but it's mainly used to connect the internet to a computer or network.

NETWORK TOPOLOGY

A network topology is the physical or logical layout of nodes and links and how they are arranged to create a network architecture. Depending on the topology, a network can be either fast or slow—reliable or unreliable. Most common network topologies include bus network, star network, tree network, ring network, mesh network, and fully connected network (see figure 9.2).

- Bus network—in this type of network the nodes are connected to the single linear network link called a bus. All the nodes exchange data through the

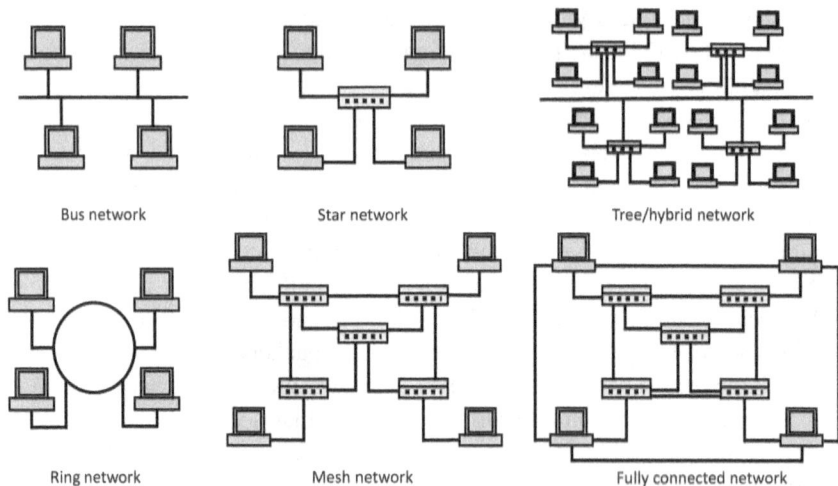

Figure 9.2 Six Common Types of Network Topologies. Author Created.

bus and each node or host in this type of network is called a station. This is a very cost-effective topology and any station can be connected to the link easily, but this type of network is prone to failures and not very reliable. A bus network can be created using coaxial cable (bus) and Bayonet Neill–Concelman (BNC) connectors for connecting the stations to the cable.
- A computer with NIC that supports the 10BASE2 Ethernet variant can only be connected with a bus network because such NICs have a BNC connector. Most of the hardware is obsolete because this type of network is not commonly used nowadays.
- Star network—this network has a star-like architecture and all the nodes are connected to one central hub. Hub is a device that works at the physical layer to connect various nodes in a network. This type of network is most commonly used and it's better than the bus network. It's expensive to implement the hardware but still more reliable than the bus network.
- This type of network requires hardware like twisted pair cable (like category 3 [cat 3] and category 5 [cat 5]) or optical fiber, registered jack 45 (RJ45)/8P8C connectors (used in 10BASE-T, 100BASE-TX, 100BASE-T4 and 1000BASE T Ethernet variants) for connecting the nodes having a NIC with similar connectors, and an Ethernet hub or switch.
- Tree network—this type of network is also called a hybrid network because it has an architecture of both star and bus networks combined all together. Here the star networks are connected to a bus forming a bus network also. This network is better than the bus network in a way that it has individual hubs only connected to the bus instead of the client computers, which

reduces the overall network traffic of the bus, but the infrastructure is more expensive than a bus network.
- The hardware requirements are similar to bus and star networks but the hub would require a BNC connection.
- Ring network—in this type of network every node is connected in series and the last node is also connected with the first node forming a circular ring-like architecture. Data travels from one node to another—each node handles and forward the packet—to deliver it to the destination node. This type of network is better than the bus network and also cost-effective because no central networking device like hub or switch is required, but problems in any node can affect the entire whole network.
- Hardware requirements for this type of network are similar to bus and star networks but hub/switch is not required.
- Mesh network—this network has a mesh like architecture and all the infrastructure nodes like networking devices (hubs and switches) are connected directly with each other forming a mesh, and other nodes like the client computers can connect to any of the infrastructure nodes to access the network. The mesh nodes work dynamically to route data efficiently and have better fault tolerance.
- This type of network is better in a way that if any node fails then it doesn't affect the entire network, but more expensive to create the infrastructure. The hardware requirements can vary depending on the infrastructure of the network and can be either wired or wireless.
- Fully connected—it's a network architecture almost similar to the mesh network, but the only difference is that in this network every infrastructure node is connected with each other and other client computers are also connected with each other. This type of network provides the highest network infrastructure and reliability. The hardware requirements can also be similar to the mesh network.

The Ethernet variants—10BASE-5, 10BASE-T, 100BASE-TX, 100BASE-T4, and 1000BASE-T—belong to IEEE 802.3 Ethernet standard for LAN, and their starting digits represent the maximum bandwidth in Mbit/sec, whereas the technology used for wireless LAN (WLAN) belongs to IEEE 802.11. Both these standards specify the protocols used for the physical layer and the data link layer's MAC for wired Ethernet and WLAN.

Chapter 10

What's the Internet?

The internet is a global system that consists of an indefinite number of computer networks that uses the transmission control protocol/internet protocol (TCP/IP) for connecting computers around the world using various networking devices. Just like the computers on a local area network (LAN) can exchange data between each other using various networking devices like switches and routers, a computer connected to the internet can also share data with any computer connected to the network from any part of the world.

However, the sharing of data over the internet is not done directly between two computers like in a LAN. A computer or any device connected to the internet is identified with an IP address and is called as a host (see figure 10.1).

To get internet access, a computer must be connected to an Internet Service Provider (ISP) like AT&T and Google Fiber. An ISP is connected to other ISPs and Point of Presence (Pop) like data centers located in various locations of the world. ISPs are interconnected globally through internet backbone networks. This layout of the internet connection is categorized into three types of networks—tier1 (internet backbone providers), tier2, and tier3.

Tier1 networks can peer with other tier1 networks globally forming an internet backbone. Tier2 networks are ISPs and PoPs that connect other ISPs (tier3) or users (business related) to the internet through the internet backbone networks. Tier3 networks are ISPs that provide the internet to all types of users (business, consumers, etc.).

Every ISP provides internet connection to computers that includes clients and servers locally, either wired or wireless. Servers are those types of computers that provide software products or services to any authorized computer connected to them. Servers can also provide storage to store data globally and are called database servers, and some can be used to provide communication

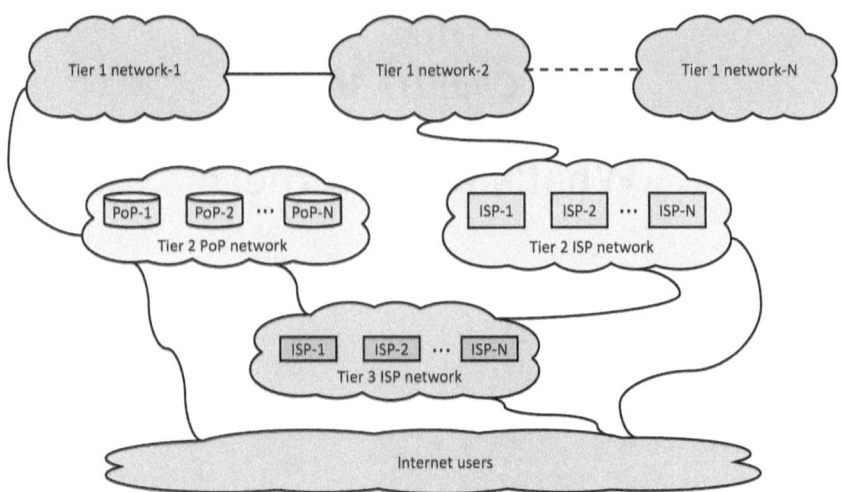

Figure 10.1 Various Tiers of Internet Providers. Author Created.

services like email, which are called mail servers. Servers are mostly popular over the internet for hosting websites and are called web servers.

A client, on the other hand, is a user's computer that gets connected with a server to access its software products or services. The client computer should have an internet connection to get access to a remote server of the internet. That's why an ISP authentication is needed to get internet access and only after successful authentication—a computer gets connected to the internet. The ISP works like a middleman between the server and a client. Because of this reason, the internet works mainly on a client-server model.

However, this client-server model has evolved into cloud networking, where indefinite servers can be connected to form a cloud network through which cloud services can be provided to the users. Many cloud service providers are connected to the internet to provide cloud services like Infrastructure-as-a-Service (IaaS), Platforms-as-a-Service (PaaS), and Software-as-a-Service (SaaS).

IaaS provides virtual infrastructure for a network or a user like backup and virtual machines. PaaS provides a platform like operating system (OS) or middleware (e.g., Java runtime, .NET runtime, integration), to develop or run applications, without the need for building or maintaining the network infrastructure. SaaS provides apps to the users that can run directly from a web browser regardless of the OS or the underlying hardware, like running the same office app on two computers with different OS.

ISPs can provide internet connection within a limited area either through wired connections like telephone lines, coaxial cables, and fiber-optic cables, or wireless connections like a mobile broadband network. The wired connection is usually done using the digital subscriber line (DSL) technologies like symmetric digital subscriber line (SDSL) and asymmetric digital subscriber line (ADSL) using telephone lines.

Data over cable service interface specification (DOCSIS) is used for providing broadband internet using coaxial cable used for cable television. Passive Optical Network (PON) fiber-optic wired connections like NG-PON2 are used to provide high-speed broadband internet connections.

Wireless internet connections are mostly provided by the mobile service providers with Long Term Evolution (LTE) connections like LTE Advanced Pro and 5G connections can provide high-speed wireless internet connections. Such mobile service providers are also ISPs and require Subscriber Identity Module (SIM) information for authentication. However, the speed of the internet can vary depending on the location of the mobile device and network signal strength.

Most of the internet connection uses the protocols included in the internet protocol suite, and every computer or networking device connected to the internet must have a unique IP address.

This IP address is usually assigned by the ISP when the internet connection is accessed through the ISP. IP addresses can either be static or dynamic—a static IP address assigned to a computer remains permanent, but a dynamic IP changes each time a new connection is made with the ISP, and it's done using Dynamic Host Control Protocol (DHCP) at the application layer.

Since the past, the internet was mostly used for information gathering through web surfing (web navigation) and communication like electronic mail (email) and instant messaging.

Web Surfing is a way of exploring websites on the World Wide Web (WWW), also known as the web. The web is a virtual place over the internet that contains various information in the form of web pages and media as content.

A web page is a type of document that contains hypertext—a type of text that can refer (hyperlink) to other text, which can be accessed with a click (using a mouse) or tap (using a touch screen). Webpages are created using Hypertext Markup Language (HTML) and cascading style sheets (CSS), but sometimes code of other programming languages like Java and Python are also used.

Such web pages are stored on a web server that can be accessed using an IP address or a uniform resource locator (URL) link with the help of a web browser like Google Chrome or Mozilla Firefox. An URL is like a web address that contains the website address and the subaddress of a particular web page.

A website address is a string of characters containing a protocol name, subdomain, domain name, and a top-level domain. The protocol name and subdomain are separated using a colon and two forward slashes. The domain name is kept between the subdomain and the top-level domain, separated by dots. For example, http://www.google.com is the website address of Google, and here the "http" is the protocol name, "www" is the subdomain, "google" is the domain name, and ".com" is the top-level domain.

The subdomain "www" represents the World Wide Web and for certain websites it's not necessary to mention it in the website address while accessing it on a web browser. Both the domain and top-level domain are together called the root domain, and any website is registered with a root domain name.

The internet has a Domain Name System (DNS) for naming any website and it's a hierarchical tree-like structure containing DNS name spaces in various zones. Any computer or device connected with the DNS system must get a registered domain name, where the subdomain and top-level domain names are given depending on their zones. The topmost zone is called the root zone and it doesn't have any name but is identified by a dot.

That's why every website address actually ends with a dot but as modern web browsers can resolve the root domain automatically, and it's not necessary to include any such dot. For example, "www.google.com." is the same as "www.google.com" for any modern browser.

Below the root zone all the top-level zones of the DNS system exist and they are named according to their type and country. For example, .com represents commercials, whereas .us represents the United States.

Similarly, subdomain also gets its name depending on the type of the zone, and it stays below the top-level domain zone. For example, a website called example.com offering various services can name its blogging site as blog.example.com.

To access a website, the website address should be entered in the address bar of a web browser and either the prompted site name should be selected or the "enter" key should be pressed on the keyboard. After the website address is entered in the web browser, it sends the request to the DNS server of an ISP. A DNS server holds the records of all the registered websites along with their root domain name and their associated IP addresses.

Most websites have a static IP address and never change. So the website address sent by the web browser is checked by the DNS server in its records and if any match is found, it transfers the connection to the server that is hosting the website. When any website is accessed using the website address (without using any subdirectory), the home page is displayed on the web browser first.

Every webpage comes with data that are temporarily stored by the web browser and are called cookies. Such cookies can store personal data like username and password locally and most websites won't load or allow access to services if cookies are not enabled in the web browser. In this way the user gets access to a website using a web browser on a computer.

Besides web surfing, the internet is well known for sending emails. Email is a method of sending messages between users connected to the internet through a computer.

Email can be sent or received using either an email app or with a web browser. An email app uses Post Office Protocol (POP) or Internet Message Access Protocol (IMAP) to send and receive emails. On the other hand, a web browser uses simple mail transfer protocol (SMPT) to send and receive emails.

Most email providers have dedicated email servers or cloud services. The email app or a browser must first connect to a mail server to access the email service. Email apps like Mozilla Thunderbird can be used for sending and receiving emails on a computer, but such apps take time to synchronize the inbox. However, emails services like Gmail can be accessed using a web browser like Google chrome without the need to synchronize, and the inbox can be accessed quickly.

To send an email to any user, an email address should be entered in the recipient's "To" field of the email interface. The email address contains two parts—first part is usually the email ID or username followed by the at sign (@), and a domain name of the mail server (e.g., gmail.com, outlook.com, aol.com). To send an email to multiple users, more than one email address (separated by comma) can be entered in the "To" field.

But the total number of emails that can be sent at once can vary depending on the email service provider, but up to one hundred is mostly allowed, otherwise it is detected as spam. It's also possible to send more than one email as a carbon or blind copy by entering the recipient's email addresses in the carbon copy (Cc) and blind carbon copy (Bcc) fields of the email interface.

Email sent to Cc recipient(s) will not see the email address(es) of the Bcc recipient(s), but the Bcc recipient(s) can. Emails can be sent as plain text or with rich text, and can also contain file attachments.

Almost every graphical interface of an email client includes a "send" button for sending the email, but other options like scheduling the email to be sent at a specific time or repeating the send operation are also available in some apps and web-based emails like Gmail.

MCI Mail was the world's first commercial email service provider in 1983 and it was operated by MCI Communications Corp. till 2003.

Along with emailing, internet users have been using various instant messaging apps and services to communicate with other users around the world. Such messaging apps use protocols that are mostly proprietary in nature. For example, the Skype messenger uses the Skype protocol for communication. Apart from messaging, most instant messengers have features like live audio/video calling, media sharing, group chat, and so on. For example, the WhatsApp messenger can be used for audio/video calling, media sharing, and group chats.

The AOL Instant Messenger (AIM) of AOL was released in 1997 and became the leading online instant messenger of the United States. Later, the Yahoo messenger (Yahoo! Pager) released in 1998 became a popular instant messenger, which was used all over the world for many years.

Besides web surfing, emailing, and instant messaging, nowadays the internet is also used for blogging, vlogging, watching online videos, social networking, online gaming, online shopping, online schooling (virtual school), online banking, and so on.

INTERNET ACCESS

To get internet access the user must have an internet connection on a working computer or any other device like a smartphone that supports the internet protocols. Internet connection is done using a communication medium, modem, and ISP (see figure 10.2).

- Communication medium—this is a type of medium through which data is transferred to establish an internet connection. It may be either a guided media like telephone lines and fiber-optic cables or unguided media like the radio waves used in a 5G network.

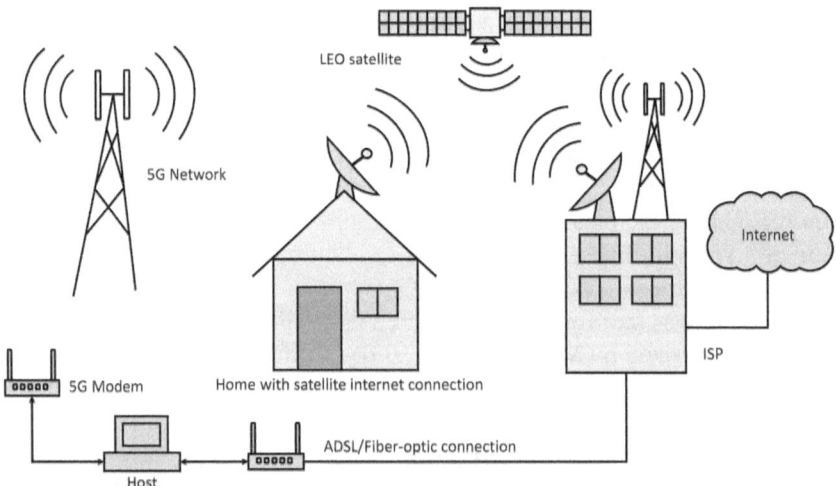

Figure 10.2 Various Types of Internet Connections. Author Created.

- Modem—it's a device that lets data communication between a computer and the server or any other device on a network or internet. There are various types of modems used for different types of internet connections. For example, ADSL modems are used to get a wired internet connection through the telephone lines, whereas, a 5G modem is used for getting a wireless internet connection through a cellular network.
- ISP—this is the company that provides internet service and it's the main access point through which a user gets internet access, and it's like a main gate to the internet world. ISPs always check for authentication before giving internet access, and the user details like username and password are used in some internet connections like ADSL for authentication.

Other mobile internet connections use the SIM details for authentication. ISPs charge for internet usage based on the monthly plans selected by the user, which is called a postpaid internet connection, but prepaid (pay-as-you-go) mobile internet connections can be used by selecting a data plan and paying before using it.

ISPs can block internet access to some websites and services, but in such situations the user can get access to a blocked website by using a virtual private network (VPN) app like HotSpot shield or using a proxy IP address and a port number with the web browser.

Internet connection speeds can vary depending on the communication medium and the connection plan selected by the user. Internet connection can be either narrowband or broadband, with narrowband being the slowest one and it was used in telephone lines for dialup connections in the past. A narrowband internet speed can be up to 144 Kbit/s only, whereas a broadband internet speed can reach up to 50 Gbit/s (for fiber-optic connection) and 2 Gbit/s (download rate) for 4G LTE cat20 mobile connection.

Apart from the mobile network, a wireless internet connection can also be done using a satellite, which is called satellite internet access. With this technology a user can get internet access directly from a geostationary or Low Earth Orbit (LEO) satellite with the help of a modem and a satellite dish antenna. The bandwidth may vary depending on the internet provider and the satellite, and the bandwidth can go over 100 Mbit/s. Such satellite internet connections are not available in all countries. However, Starlink (SpaceX) is one such satellite service provider that is providing internet services in many countries.

There are also portable satellite devices like satellite modems and phones through which internet access is possible but the bandwidth is not high and it's expensive too.

Even though ARPANET was the first wide area network (WAN) to successfully run a computer network in the 1970s, the internet backbone was created by the National Foundation Network (NSFNET) in 1985. An internet backbone is a tier1 network that connects ISPs and other tier1 networks internationally. Because of an internet backbone the internet can expand globally, and after the ARPANET and NSFNET decommissioned in the 1990s—many other modern internet backbone providers like AT&T, Verizon, Sprint, and Lumen started connecting the ISPs globally.

The Internet works mainly with the help of TCP/IP protocol, which would not have been possible without the protocol—thanks to Vint Cerf and Bob Kahn, for developing it in 1974.

INTERNET SECURITY

The internet is a place where a user can face many kinds of security issues. The most common types of internet security concerns include malware, hacking attacks, phishing, and spamming.

- Malware—it's a computer program designed to cause harm to the computer or the user by getting access to the user data or taking control of the system. Computer viruses are a kind of malware that can even steal sensitive data like username/password or any online banking details like the credit card number. Viruses like Trojan horse and ransomware can not only take control of the computer and steal confidential data, it can also restrict access to the stored data and can ask for a ransom payment in return for accessing the data.
- Other malware like spyware can keep track of the internet activities and can send the browsing history to servers. Spyware can even use other hardware devices to monitor the user's activities.
- Such malware can enter the computer while visiting any untrusted website or can come with any video game or app—downloaded from the internet.
- Hacking attack—hacking is a process of accessing someone's computer or online account and also taking control of a computer, without the consent of the user. Hacking is done using hacking tools and techniques. A person who hacks or tries to hack a computer is called a hacker. Hackers usually hack a computer (over the internet) anonymously by concealing the geolocation and the original IP address.

This is done using a proxy server or an anonymous network like Tor. Hackers can intrude on someone's computer by knowing its IP address and the port number, which they find out using various packet analysis techniques. Other methods like brute force are also used but not always successful. Hackers also use a technique called dictionary attack for cracking usernames or passwords.

Some hackers can use A Man-In-The-Middle (MITM) attack to obtain sensitive information. This is done by sending a link to a webpage (clone) that would look similar to the one that the user uses to login. And when the user would enter the login details, it would be sent to the hacker instead. MITM is done through Wi-Fi connections also, and any user using a Wi-Fi router can encounter an MITM attack.

- Phishing and spamming—this is the most common issue for email users, who can suffer from phishing and spamming emails. Phishing is a process of sending emails that would contain a fake message, which would trick the user to reveal sensitive information, and some messages may contain a link that would redirect the user to a fake website, which would ask the user to enter sensitive information to get something rewarding in return.

Some phishing messages could even come with attachments that would contain a computer virus.

On the other hand, spamming is the process of sending the same email to too many recipients repeatedly for any commercial (advertising) or any profitable purposes. This is usually done with the help of some spamming software, but nowadays most spam filters don't allow spamming.

- Phishing and spamming can also happen with any other messaging services like WhatsApp or any other websites that offer online messaging interfaces.

To avoid any such internet security problems, the user should use a good antivirus app with built-in internet security features. Avast is a popular antivirus that comes with the best internet security features like a firewall and hacking protection. Apart from Avast, there are many other antivirus apps that can be installed on a computer for internet security. The top antiviruses as per their performance can be found at www.av-test.org (AV-TEST), which is an organization that tests and rates antivirus and internet security software as per their criteria.

The internet has also become a place for content and software piracy, and users can download copyrighted content using various download managers and peer-to-peer file-sharing apps, which has become a big concern for the content creators and software companies.

Chapter 11

How to Choose a Computer

Choosing a computer is like selecting the right vehicle for transportation needs. There are many types of vehicles starting from family cars to buses and trucks, but every vehicle is designed to work in certain environments and to take specific types of loads. To travel short distances within a city an Electric Vehicle (EV) can be used, but to travel far-off places on bad roads, a Sport Utility Vehicle (SUV) like Ford Explorer can be useful. Similarly, to choose a computer for any work, the requirements need to be known, and then select the type of computer that would meet the requirements.

This process of selecting a computer requires a few steps of logical decisions that need to be done carefully after knowing the purpose of a computer. Whether it's about buying or building a computer, the hardware configuration could vary depending on the purpose of use. Even the software requirements could vary depending on the work and hardware configuration.

For example, a computer with Intel Celeron processor won't be suitable for productive work and playing high-definition games like Forza Horizon. On the other hand, a computer with Intel Core i7 processor and a high-end graphics card will be a waste if it's used only for web browsing or sending emails.

That's why choosing the right computer can not only make things easygoing but can also give a better value for money. But selecting the computer could be sometimes challenging and might become an endless pursuit for that dream computer. However, this may not become too complicated if it's done with the right method.

There may be many ways of selecting a computer, but the simplest way to do this involves mainly two steps—know the requirements and prepare the system checklist (see figure 11.1).

After completing these two steps—it will become easy to either buy or build a computer. The first step is the most important one, because knowing

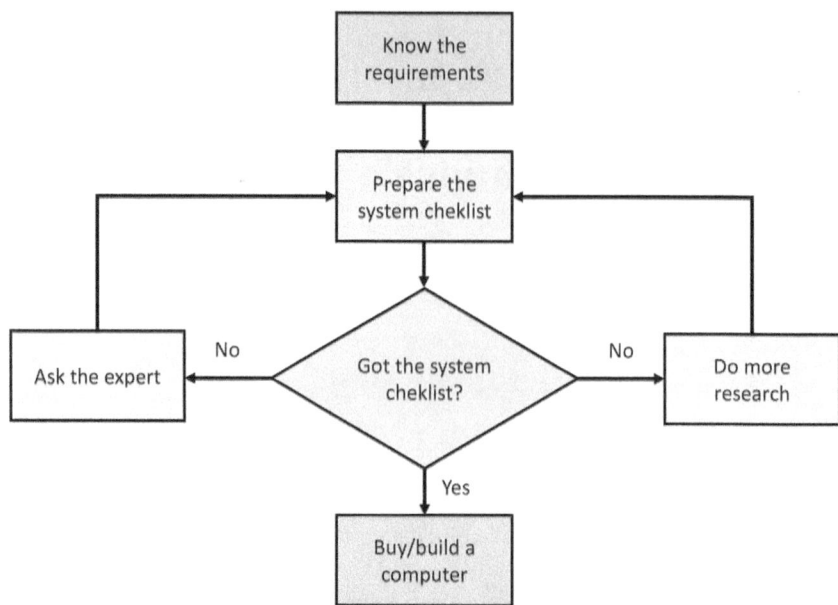

Figure 11.1 Computer Buying Process Flowchart. Author Created.

the requirements for using a computer will only determine which type of computer will be suitable for use. After knowing the requirement, the next step will be to prepare a system checklist (hardware configuration and the required software). It can be done either by doing some research or taking help from a computer professional. After preparing the checklist successfully, a computer can be either custom built (using hardware parts) or a pre-built one be purchased from the manufacturer.

KNOW THE REQUIREMENTS

The most important thing before choosing a computer is to know the requirements and the purpose of using a computer. A computer is a machine that has hardware and software. The hardware is like the car and software is the driver. The ability of a car to carry load and how fast it will travel will depend on the mechanical parts and the engine. But how it will move on a road and which direction it should go will depend on the driving skills of the driver.

An EV like Tesla is good for traveling on road with an average speed of a family car, whereas an SUV can not only travel on road but also can go off-road. That's why a vehicle can be selected based on the travel destination and a driver who is familiar with the route and the vehicle will also be required.

Similarly, the hardware of a computer is like a car, and the ability of a computer to do certain tasks will depend on the hardware. On the other hand, the software is like a driver and it will tell the computer how to perform and how to do the task. That's why, first the requirements of using a computer have to be understood. This will also give an idea about the hardware configuration and the software like operating system (OS) and apps that need to be installed on a computer.

The hardware configuration can vary depending on the purpose of using a computer. For example, an Intel Celeron-based computer without any discrete graphics card can be used only for study and office work like web browsing and document editing. This type of a computer is also called an everyday computer, which is of course a low-end computer. On the other hand, a computer with AMD Threadripper processor and a discrete graphics card can be used as a workstation for productive work like product design using Computer Aided Design (CAD) app or for video editing purposes, using a video editing app.

PREPARE THE SYSTEM CHECKLIST

To buy or build a computer, a system checklist should be prepared, which will include the hardware configuration details and the software requirements. To prepare the hardware configuration, the motherboard should be selected first. This is because almost every motherboard contains a chipset with a chipset number. Not all chipset can support all types of processors (CPU). For example, two motherboards—one with Intel chipsets H410 and another with H510—can have a LGA 1200 socket, but only the one with H510 chipset can support an Intel rocketlake processor.

The size of a motherboard also matters, and smaller motherboards like mini-ITX can be used for a computer that will perform between a low-end and a high-end operation. And it will also consume less power than a high-end desktop computer. It's mostly seen that smaller a computer is, less is its power consumption, but smaller computers with high-performance hardware are always expensive.

Small-sized computers are good for traveling purposes and also convenient for placing on a desk where space is limited.

A medium-size computer can have a microATX motherboard, and this type of motherboard is somewhat bigger than a mini-ITX, but smaller than a regular ATX motherboard. Desktops with microATX may not be as suitable as the mini-ITX for traveling, but can be somewhat portable.

Computers with a microATX motherboard may be configured to work like a high-end computer, but it might still have some hardware limitations and may not include all kinds of hardware components that a regular ATX motherboard can have. For example, most microATX motherboards can have only two DIMM slots for the RAM modules. Whereas, almost most recent ATX motherboards have four DIMM slots and can support dual channel memory.

A desktop motherboard is usually of two types depending on the processor manufacturer. As most desktop processors are manufactured either by Intel or AMD, motherboard manufacturers also manufacture the two types of motherboards. A motherboard that supports AMD processors can't support Intel processors—neither the motherboard that supports Intel processors can support AMD processors. That's why this is also very important to choose whether the computer should have an AMD or Intel processor.

Generally, it's seen that AMD processors are better for media processing and virtualization because of having more cores. That's why AMDs are best suitable for workstations and servers. Intel processors are priced bit lesser than AMD and can be used at consumer-level computers. However, for gaming both AMD and Intel seem good and there is no difference in value for money.

It's also important to check how many PCIe slots are present, and it's also possible to install more than one graphics card on a motherboard with more than one PCIe slot. This can improve the graphics processing performance—required for running games and simulation apps.

Apart from the PCIe ports, the number of SATA ports should be checked as well and more than one SATA port can enable connecting more storage drives, and ensure that it's the latest SATA III version. Even some motherboards come with M.2 slots for M.2 SSDs, and using M.2 can greatly improve the performance of a computer.

The I/O interface of a motherboard is also a very important thing to check for. The I/O should contain enough USB ports and should support the latest USB 3.1 version. Other types of ports like HDMI ports may be present but it may not be required when using a graphics card. Audio ports are also included with the I/O interface and some other ports like Ethernet may also be present. But the USB and audio ports are the most essential I/O ports.

Some motherboards come with a processor and such motherboards can't be upgraded because the processor is soldered to the motherboard through a BGA socket. Before choosing such a motherboard the processor specifications should be checked for usability.

After selecting the motherboard, the processor should be selected. If it's an AMD processor, then it should be decided whether to go for a Ryzen family processor or a Threadripper (if supported by the selected motherboard). This decision should be based on the performance requirements of the computer and the budget. Ryzen processors start from Ryzen 3 and all the way up to Ryzen 9. Greater the number, the more powerful the processor is.

In case of Threadripper processors, they come as regular Threadripper processors with a model number and also as professional Threadripper processors (Threadripper PRO) with model numbers. Ryzen processors can be used for any low-end to high-end computers, but Threadripper processors can be used for high-end computers only, where the Threadripper PRO is mainly used for workstations.

Intel processors, on the other hand, come with different types of processors but could belong to the same family. The most low-performance processors are called Celeron and better than that are the Pentium series. But the most exciting line of Intel processors starts from Core i3 to Core i9. A Celeron processor should be chosen for a low-end computer and any Pentium processors having dual cores (with hyper-threading) can give performance better than a Celeron, but may not perform like a high-end computer.

Intel Core i3 processors can give mixed performance results depending on the model number and some may not perform any differently from the Pentium processors. However, processors starting from Core i5 could perform better than the Core i3. Intel Core i7 and i9 can be used for a high-end computer and they are mostly used for productive work like video editing and gaming.

The Intel Xeon processors are used in some workstations because of having too many cores, but Core i9 can also be used.

A discrete graphics card should be used only if the computer is used for gaming and productive work like product design using CAD or video editing. Nvidia is the most preferred graphics card of all time, but the AMD Radeon graphics cards can also be used. The main thing to check in a graphics card is the maximum supported frame rate (fps) and the memory. For gaming purposes a graphics card with 120 fps and 8 GB memory specifications are always the best, and can also be used for productive work.

The random-access memory (RAM) capacity should be decided based on the purpose of use, and for a low-end computer 4–8 GB RAM is enough. But for a high-end computer the RAM should be 16–32 GB or more. That's why while selecting a high-end computer, the maximum supported RAM capacity of the motherboard should be checked, so that the RAM can be upgraded as per the requirements. Such RAM upgradability can make a computer upgradable in the future and can give better value for money.

Just like RAM, storage devices like HDD and SSD are also required to install the OS and data. To make the performance of the computer better and reduce the startup time, it's always better to install the OS on a SSD instead of an HDD. This is because HDDs have mechanical moving parts and take more time to respond than the SSD (which don't have any mechanical parts). M.2 is one such SSD that can be used to improve the performance.

Using such an SSD with a low-end computer can also improve the performance and startup time. As SSDs are more expensive than HDDs, SSDs like M.2 are used only to store the OS, because OS like Windows 11 can take 64 GB of storage space, and any 128 GB SSD can be afforded to store only the OS. On the other hand, an HDD is cheaper than the SSD, and can be used to store large data like videos. That's why vloggers and gamers use HDDs with higher capacities like 5 TB to store videos and HD games.

Other peripheral devices like display, keyboard, and mouse should also be selected based on the requirements. For example, the size and type of the display will depend on the purpose of use. A simple 22" TFT LED display (16:9) is enough to do everyday work on a computer, but using IPS with a supported frame rate of 120 fps can be used for gaming and video editing.

A keyboard can be simple or a gaming one with backlit features, but spending more on a keyboard won't be useful unless such a requirement for a high-end keyboard is there. A wireless keyboard can also be useful to get rid of the cable but ensure that it has a power switch to turn off when not in use, otherwise it will run out of battery frequently. A mouse can also be wired or wireless and the same things need to be checked. Some gaming mice can give extra buttons for advanced functionality but it will be only useful for a gamer.

Apart from all these hardware, the software should also be selected. Depending on the purpose of use, the software can vary. Software may include the OS and third-party apps. The OS should be selected based on the requirements and the hardware configuration. A low-end computer with a 1 GHz dual core processor can support Windows 11 but whether to use Windows 11 Home or Pro version will depend on the purpose of use.

Whether it's a low-end or a high-end computer, if it's used for business then Windows Pro should be used because Windows 11 Home does not have the features that are required for any business use. Another alternative to Windows would be using the Ubuntu OS, but it may not run properly on a low-end computer, and not everyone can find it easy to use like Windows.

Different apps can be either downloaded for free or can be purchased from the manufacturer to install on the computer. Microsoft Office is the most commonly used app suite that needs to be installed on a computer (whether it's a low-end or a high-end one).

This app suite comes with a word processor called MS-Word, a spreadsheet app called MS Excel, and a bunch of other office apps. This app suite is not only useful for office work but also for doing homework and other educational purposes.

Besides that, other multimedia apps like media players, web browsers, and antiviruses can be downloaded for free. Some professional apps used for productivity like Adobe Premiere Pro, Adobe Photoshop, Fusion 360, and so on, need to be purchased.

The following list includes the important hardware gear that should be included in the system checklist:

- Chipset/motherboard
- Processor
- Graphics card (required for a high-end computer)
- RAM capacity
- Storage devices
- Power supply unit (PSU)
- Computer case
- All-In-One (AIO) cooling system (required for a high-end computer)
- Display
- Keyboard and mouse
- OS and apps

The following list is an example of a low-end computer's hardware configuration:

- B250/Gigabyte GA-B250-HD3P
- Intel Celeron G3920
- Samsung 8 GB DDR4 3200 MHz RAM module
- Hynix 128 GB SSD and WD 1TB HDD
- Coolmax V-500 Series 500W (PSU)
- Fractal Design Meshify C Computer Case
- LG 22" TFT LED display
- Logitech wireless keyboard and mouse (combo)
- Windows 11 Home
- MS Office
- Freeware (media player, web browser, antivirus, etc.)

Purpose:

- Everyday use
- Education
- Small business

The following list is an example of a high-end computer's hardware configuration:

- WRX80/ASRock WRX80 creator
- AMD Ryzen Threadripper Pro 5995WX
- AMD Radeon RX 6950 XT

- Samsung 128 GB DDR4 3200 MHz RAM module
- Hynix 256 GB SSD and WD 1TB HDD
- EVGA SuperNOVA 1000 G5, 80 Plus Gold 1000W (PSU)
- ASUS TUF Gaming GT501 Mid-Tower Computer Case
- Vetroo V240 Black CPU Liquid Cooler
- LG 27" UltraGear 4K IPS display
- Logitech wireless keyboard and mouse (combo)
- Windows 11 Pro
- MS Office
- Freeware (media player, web browser, antivirus, etc.)
- Adobe Premiere Pro, Adobe Photoshop, Fusion 360, etc.

Purpose:

- Content creation
- Product design
- Software/game development
- Business
- Gaming

The above checklist includes the hardware specifications (not the brand) that could be configured online on the websites of some computer manufacturers like Dell, before buying. However, only the OS might come along the computer but the rest of the apps need to be installed separately.

Apple iMac can only be purchased from the Apple store or from the website of Apple, but can't be custom built and it's also not much configurable. Moreover, the iMac can only be used for productivity and business purposes because its configuration is always better than any Windows low-end computer.

The system checklist can be prepared either by doing research online or taking help from a computer professional.

After preparing the checklist, a computer can either be bought from a manufacturer or can be custom built using tools and buying the hardware parts from the manufacturer. But to build a computer, a desktop computer case (along with a power supply unit) will also be required.

Chapter 12

How to Buy a Computer

Whether it's about buying a pre-built computer or buying parts for building a computer, the only thing that matters is getting it at the right price within the budget. Buying a pre-built computer could be a bit more expensive than the one that could be built with the parts bought from a store. However, a pre-built computer can be of good quality with excellent configuration if it's purchased from a good manufacturer like Dell or HP. Moreover, buying a pre-built computer can save the time that would take for building a computer and is also a good option for those who are not capable of building a computer.

Pre-built computers and hardware parts can be purchased either from any offline store like Walmart, BestBuy, and Microcenter or from any online store listed on Amazon and eBay. Prices of computer and hardware parts can vary depending on the store. Sometimes the same type of computer or hardware parts may be priced high on an offline store, but may be priced low on an online store, and could also happen the either way.

That's why it's better to first step-in an offline store to check the price of the computer or the parts. If not sure about which store to choose, Walmart could be the best option to get a pre-built computer or any hardware parts at the best price. Other stores like BestBuy and Microcenter can also give a variety of options. Microcenter also has a build your own (BYO) section for a computer to be custom built after buying the hardware parts.

Before buying a computer or the parts, first check the price tags of the computer or parts (included in the system checklist) and then compare it with the price listed on other online websites like Amazon or Walmart. Go for only those items which are priced less at the offline stores. Try to visit more than two stores to get the best price and different brand options. Select a brand that can give the best price, but always choose a pre-built computer from a renowned brand like Apple, HP, or Dell. Pre-built computers should consist of good quality hardware, with a good cooling system and a power supply.

Most top-brand manufacturers pre-build computers at their state-of-the-art manufacturing facility with excellent quality control checks (see figure 12.1).

OFFLINE BUYING

When buying a computer or its parts from a retail store like Walmart, ensure that the system checklist is ready, and a few brands in mind that can be checked as well. After stepping in the store, try to check for the best brands like Apple (if planning to buy an iMac). However, it's best to buy an Apple computer from an authorized Apple store only. Just compare the prices of every brand available in the store with the required hardware configuration present in the system checklist.

In case of buying hardware parts, first start with the motherboard, processor, and graphics card. After getting the best price for these three items, the rest of the parts can be checked.

As the overall value of a computer depends on the motherboard, processor, and graphics card, these three parts are the most expensive parts of a computer, and choosing them is the first challenging thing to do. The most important thing to keep in mind is that these three parts don't come with many different brands (even if motherboards come with different brands, only a few are good) and don't give much price options on the basis of brands. As the motherboard is the most valuable part of a computer, its quality should not be

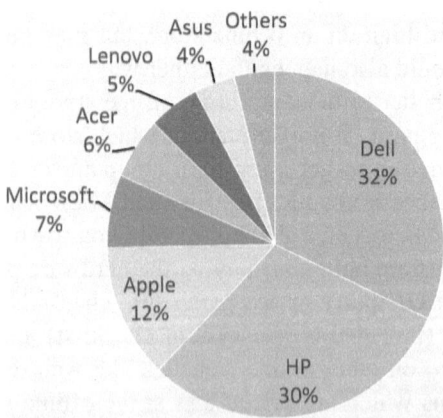

Figure 12.1 Desktop Computer Users by Brands in the United States as of 2021. Author Created.

compromised for a low budget. Always buy a good quality motherboard from a renowned brand like Intel, Asus, or ASRock.

It's better to go for a computer or the parts that have a long warranty period like two years or more. Longer warranty period means more durability and less possibility of defects or problems after use. Some companies provide the option to purchase an accidental damage protection plan and also let the customer extend the warranty period. Buying a computer with such a protection plan and extended warranty can save money for any servicing for many years.

One should finalize a computer before buying, only after visiting a few stores and comparing their offered price for the same hardware configuration.

However, hardware parts can be selected in the same store if the offered price is less than the price offered by any online store. Try to pick the parts that are manufactured or imported recently and know that by checking the manufacturing or imported date mentioned on the packaging of the product. This is recommended because most hardware parts are updated by the manufacturers and any bugs or flaws of the previous versions are corrected in the later versions.

For example, a motherboard released in the previous year might be having a chip or any component with some quality issue but the one released in the next year might come with better components and updated firmware, which would obviously improve its performance and reliability.

While buying the hardware parts ensure that the packaging has a factory seal. Some manufacturers also put a hologram on the packaging as a proof of being a genuine product, but most counterfeit products don't.

Some stores can give discounts on specific brands, but don't be tempted by that because the quality of some brands may not be that good when compared to other top brands—especially while buying a pre-built computer. There are many Chinese brands of motherboards and graphics cards that are sold at an attractive price tag, but the quality of such products may not be good and there could be even difficulty finding a service center for such products, after the warranty expires.

ONLINE BUYING

Buying a computer or its parts from an online store could save a lot of money. But that should be done only when a product is either not available or too expensive to buy it from any offline store. However, there are some issues while buying a computer or hardware parts online. Sometimes a defective or a damaged product might be delivered after buying it online. Even a wrong product gets delivered in some situations. That's why it's very important to select a reliable e-commerce company like Amazon or eBay before placing any order online. It's also very important to read the return and refund policy of the company.

Some companies have two options for any delivered items—refund on return or replacement on return. For some items full refund is issued after the item is returned back to the seller, but for some items only replacement is done. Amazon offers replacement-only option for most electronic products and that's why ensure to finalize properly about the brand, because after finding any quality issue, there will be only an option to replace it.

There are some products that are neither returnable nor refundable and that's why the return policy of the item should be checked properly before placing an order. Most computer hardware products are available on Amazon at a reasonable price but the same products are also available in the stores of other online competitors like BestBuy and Walmart. But the delivery service of the company could vary depending on the location. Amazon provides the best delivery service and also has its own logistics around the world.

Amazon also provides pickup service for the returnable items but some companies like eBay don't. If any product is bought from eBay, and for any reason if it needs to be returned, then the buyer has to return it through any shipping company by paying the shipping charges. However, Amazon does the pickup for free and also refunds the full amount without much delay.

This is why buying a computer or the hardware parts from Amazon could be a better option, but be careful while choosing the seller.

Not all sellers are good at Amazon, nor their items. Some sellers might send a defective or a similar-looking product, and it could become difficult sometimes to claim for the refund if it can't be established that a different product was sent. That's why it's important to check the reviews of the product and the seller before buying it. Amazon has an option to select different sellers of the same product and choosing a seller with good ratings can give better buying experience.

Both Amazon and eBay are good places for buying computers and hardware parts at a reasonable price, and also can be helpful while buying with a low budget in mind. On Amazon and eBay, some refurbished computers and parts are also sold at a cheaper price. Even some used computers are also sold at less than half the price of a new one. This might be a good option for those who don't have the budget to buy a new computer and most students can go for this option.

Although refurbished computers and parts come with a limited warranty period of only a few weeks or months, there is no assurance about the quality of a used computer or any hardware parts.

Before placing any order on any such online store the proper method of selecting the right product should be followed. This method is a standard way to search any product on any online store, and pay for it with minimum buying risk. However, this method is suitable only for placing orders on any website of the online store, but not while using the app. This is because the website gets loaded on a web browser with a different layout than the app on a smartphone. Moreover, while navigating any website using a computer like a laptop, more items on the website would appear on the screen, which will be helpful to compare.

First, go to the website of the online store and type in the name of the computer or the part that is required to buy and hit "enter" or click the search button. From the search list try to look for only the items that have ratings and reviews. Right-click on the products with highest ratings and reviews, and open in a new tab. Do the same for at least two or more items and then change the sorting option by selecting the price from low to high. From the search list select the items with highest reviews and rating, and also open them in the new tabs.

After opening the items in new tabs, compare their price and reviews. Pick the item that has a reasonable price but with average to best reviews and ratings. Also check the return and refund policy of the item. If not much information is found about the refund policy, then contact the customer care to get more information about the item. In case of buying too many items for the same product, try to buy only one piece of any item for the first time. And only after receiving and using the item, buy more of it later. Doing this could give some understanding about the reliability of the product and the seller.

Besides buying from Amazon or any other online seller, a computer can be purchased directly from the manufacturer by placing orders through their website. Companies like Apple, HP, and Dell provide an option to configure the hardware before placing the order, and they also provide free delivery or return service.

Whether it's about buying a computer or any hardware parts—buying from an offline or an online store—the main thing that matters is getting a good quality product at a reasonable price. Offline stores are good for buying branded products and are a good place to check things physically. In most offline stores it's even possible to get the demo of a pre-built computer and it's possible to find out about its features, which is not possible using any online store. On the other hand, most online stores give better price tags and delivery options than offline stores.

Therefore, both offline and online stores can be visited for buying a best computer or any hardware parts but it all depends on the price and availability of the products.

Chapter 13

How to Build a Computer

Building a computer could be as easy as making a cup of coffee, but it needs to be done carefully using proper tools and hardware parts. Failure to follow the correct assembling procedure could damage the hardware parts or even cause the computer to malfunction. That's why ensure that the hardware parts bought for building the computer are compatible and functional. Refer to the manufacturer's instructions before handling any of the hardware products.

For building the computer, basically three things are required—the hardware parts, an ESD safe place, and the tools required for building the computer.

Before building a computer, make a place ready for storing the hardware parts bought from the stores. The same place can be used for storing the brand boxes that came with the parts. This requirement is only for a few days (depending on the seller's return policy). In case any problem occurs with any of the hardware parts, it should be packed in the original brand box and can be returned to the seller for a replacement.

An empty desk can be a perfect place that can be used for assembling the hardware parts. Ensure that the room has enough lighting and has an AC outlet close enough to test the computer after building it. The surface of the desk should be free from any lying papers, plastic covers, synthetic clothes, and so on, because such things can cause ESD and could damage the hardware components while assembling. That's why a clean and dry desk surface is best for assembling a computer.

Don't keep any food or drinks on the desk while assembling—doing so could cause accidental spills on the hardware, resulting in damage and self-servicing.

Before starting with the build process it's important to know about the motherboard and its components. The components of a motherboard may vary depending on its size, generation, and manufacturer. Some of the components like CPU sockets, PCIe slots, DIMM slots, SATA connectors, and PSU connectors (8-pin for CPU and 24-pin for motherboard) may be found in most motherboards (see figure 13.1).

The build processes of a desktop computer will depend on its hardware configuration and the parts that are going to be installed, but the initial process of assembling the computer case and the motherboard could be the same (see figures 13.2–13.4).

TOOLS REQUIRED FOR BUILDING A COMPUTER

To build a computer not only the hardware parts are required but also the tools for building it. The following tools and accessories are required for building a computer:

- #1 Phillips screwdriver (preferably 10" long with rubber handle)
- Wire cutter
- Zip ties
- Magnetic screw holder
- Isopropyl alcohol (optional)
- Paper towel (optional)
- Flashlight (optional)
- Anti-static wrist strap (optional)

The #1 Phillips screwdriver is the most important tool on the list, and without this tool the assembling of the hardware parts is not possible. There

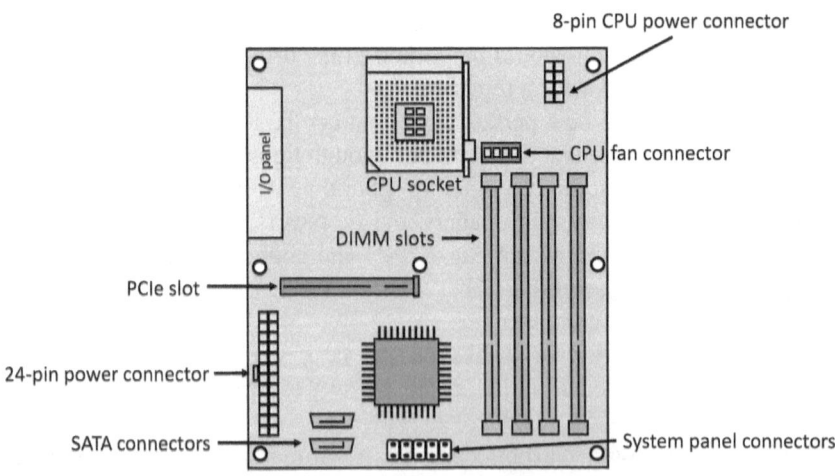

Figure 13.1 Components of a Motherboard. Author Created.

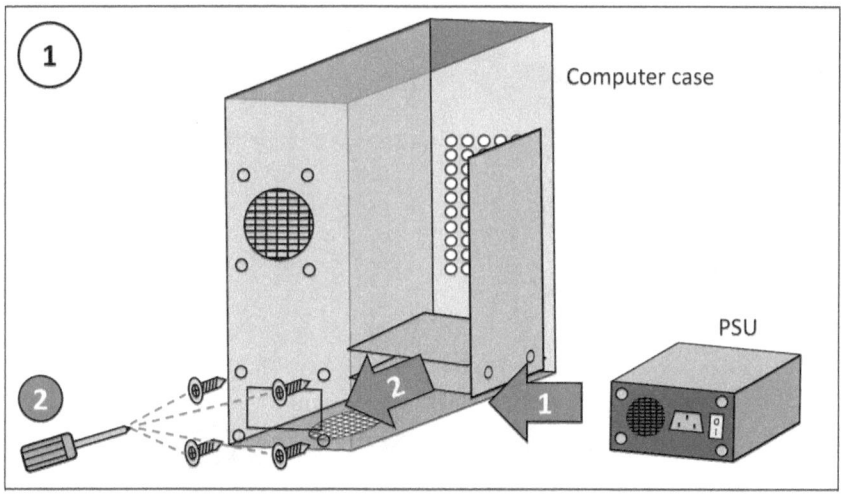

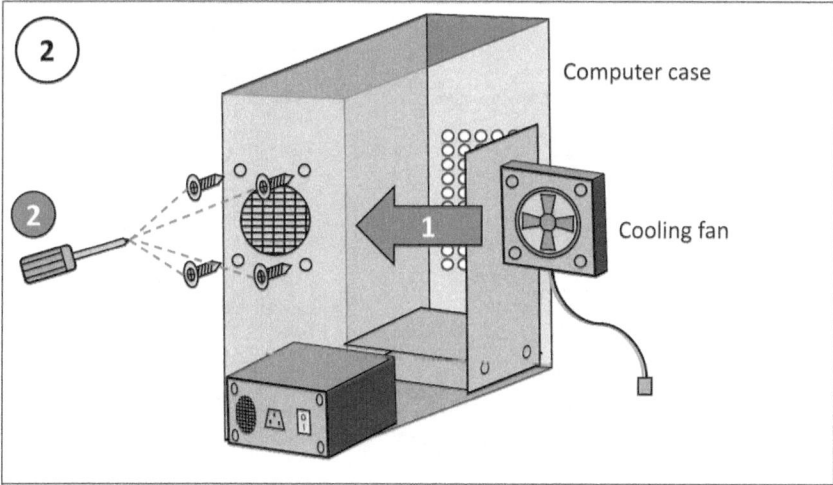

Figure 13.2 Initial Process of Assembling a Desktop Computer (Part 1). Author Created.

are some screwdriver sets that come with various tip sizes, but the #1 Phillips screwdriver works on almost every type of screw used in building a desktop computer. A battery-powered screwdriver can also be used but that will be expensive and only required by computer professionals or while building too many computers.

A standard screwdriver with a long shank (at least 10 inches long) and with a magnetic tip is useful because with a short screwdriver it could be difficult to reach some deep areas inside the computer case while assembling it. A magnetic tip can help pick up the screws while placing or removing from the

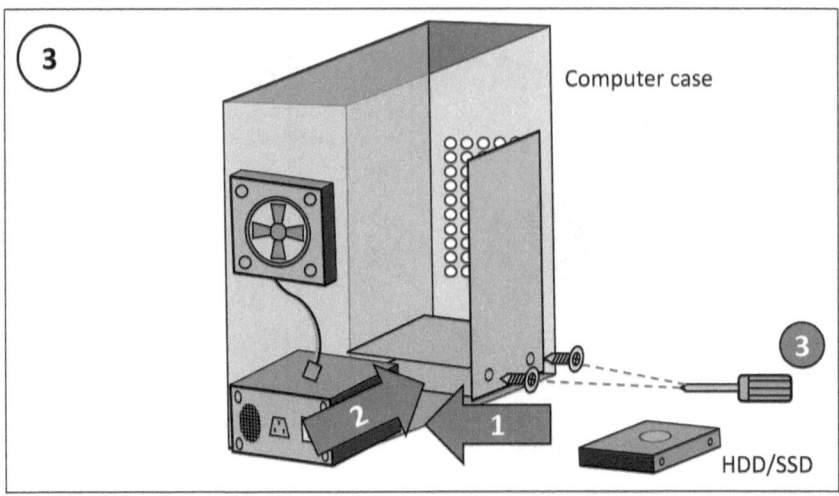

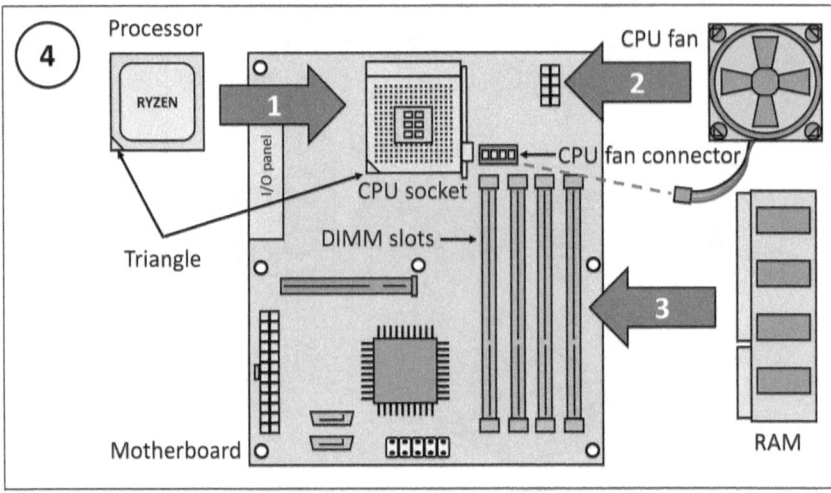

Figure 13.3 Initial Process of Assembling a Desktop Computer (Part 2). Author Created.

parts and can prevent them from falling on to other hardware parts like the motherboard.

A wire cutter will be required to cut the excess part of the zip ties, and also for cutting some wires for modified connections. Zip ties can be used to tie the wires inside the computer case after assembling, which will keep them well organized and will also look tidy. Sometimes some extra wires may be needed to provide power to some parts like fans or LEDs (if their cable is not long enough to reach the power source), and in such situations, an extra wire can be connected by stripping and cutting with the help of a wire cutter.

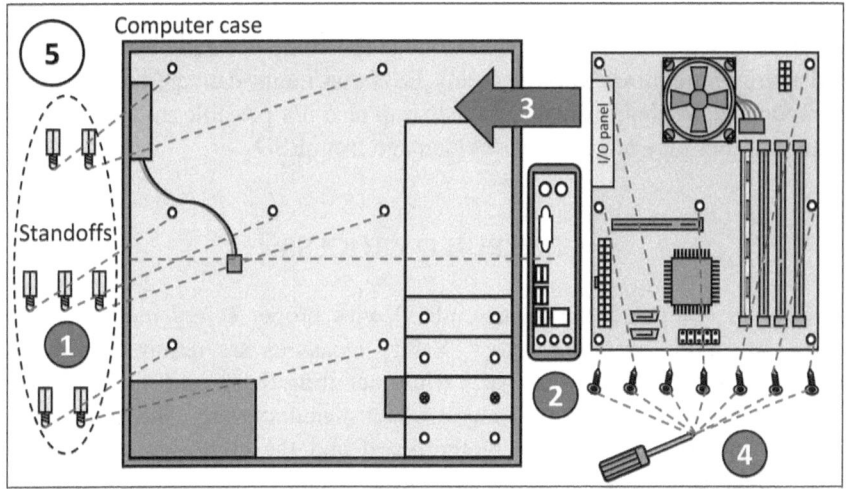

Initial Process of Assembling a Desktop Computer:

① Install the PSU.

② Install the cooling fan(s).

③ Install the HDD(s)/SSD(s).

④ Install the processor, CPU fan and RAM.

⑤ Install the motherboard inside the computer case.

Figure 13.4 Initial Process of Assembling a Desktop Computer (Part 3). Author Created.

A magnetic screw holder is a small metallic pot that has magnetic properties and is capable of holding magnetic screws tightly together in one place. This is helpful for keeping all the screws required while building the computer. But in case a magnetic screw holder is not available, any cap of a bottle or lid of a container can be used instead.

Isopropyl alcohol and paper towel might be needed to clean the surfaces of a processor and a CPU fan or the cold plate of an AIO cooler.

Apart from these, there are two things that might help while building a computer but are optional: a flash light and an anti-static wrist strap. A flash light may be required if there is not enough light in the assembling environment or to see things in deep areas inside the computer case. An anti-static wrist strap is required to prevent any ESD that might damage the hardware components. Without using the wrist strap also it's possible to build a computer but that may not prevent any damage from ESD.

THE BUILD PROCESS

A desktop computer can be assembled with proper safety measures and a few standard assembling steps. Safety measures are required to avoid any damage to the hardware parts while handling them, and it's similar to maintaining quality control of any product manufacturing. The assembling process includes making the motherboard and the computer case ready. After making them ready, the motherboard needs to be installed inside the case and all the wired connections need to be done. After all that is done, next comes the exciting step—to power on the computer and see the splash screen.

SAFETY PRECAUTIONS

Before unboxing any hardware parts, ensure that the hands are clean and dry. The surface of the desk should also be clean and dry. Ensure that there are no papers lying on the desk because depending on moisture level, a paper can develop harmful ESD. Similarly, any desk covers made from a material that is not ESD safe should be removed. In other ways, an anti-static mat can be placed over the desk before starting the build process. Most modern desktop computer hardware doesn't easily get damaged by ESD, but still there are chances and that's why it's better to make everything ESD safe before building the computer.

Ensure that there is no food or drinks on the desk and the same should not be consumed while building the computer; otherwise, there are chances of accidental dropping of food or spilling of liquid on the hardware parts while building the computer. Any food or drink contains conductive molecules that could cause ESD damage.

If there is a humidity controller (present is most air conditioners), then set it to a higher level because low humidity in the air causes ESD. Wear an anti-static wrist strap if possible, but if that's not available, then try not touching the exposed metallic surface of the computer case before handing any hardware parts. Touching any big metallic objects can discharge the static electricity from hands.

Use a magnetic screw holder to keep the screws that came with the computer case and the hardware so that they won't go missing anywhere.

MAKING THE CASE READY

The computer case needs to be made ready before installing the motherboard. To make the case suitable for use, it should have the PSU, the cooling fans, and AIO cooling system (for a high-end computer) installed in the case.

Installing the PSU

The Power Supply Unit (PSU) can be either mounted at the bottom of the case or at the top depending on the size and feasibility of the computer case. In a bigger-sized ATX case, it's better to mount the PSU at the bottom of the case (with the air intake side of the PSU facing down) because that will keep the PSU cool by pulling cool air from the bottom.

In smaller-sized cases like microATX, the PSU should be mounted inside the top back side of the case with the intake side facing downward.

To mount the PSU—first, remove the case cover and place the PSU inside the case, and push it as far as it can go toward the back of the case. Then use the screws (that came with the PSU or the case) to secure the PSU in place by tightening them using a screwdriver. Don't place and tighten each and every screw one by one—tighten the screws only after placing them properly in place and screwing them firmly first.

Installing the Cooling Fans

A bigger ATX case would need adequate cooling and at least one fan should be mounted inside the case at the back with the intake side facing inside. The case should have a positive air pressure inside, which means that there should be more fans pushing air into the case than pulling it out. To maintain a positive pressure, at least two to three cooling fans (for air intake) should be mounted at the front and one fan (for air exhaust) should be at the back side.

Some computer cases may come with only one cooling fan at the back and could cause negative pressure without additional fans. In such cases extra fans should be mounted at the front to intake more air.

Installing Storage Drives

Install any storage drives like HDDs or SSDs inside the computer case by sliding the HDD or the SSD into the mounting bracket (found inside the case) and put screws into place and tighten them firmly. Place the bracket back into the drive bay of the computer case. This process is the same for both 3.5" and 2.5" drives. Except the storage drives, any DVD drive can also be installed by sliding into the drive bay present at the top of the case. However, some cases may not come with a DVD drive bay.

MAKING THE MOTHERBOARD READY

The most important step for building a desktop computer is to make the motherboard ready before installing it in the computer case. First, take the motherboard out of the brand box and keep all the components safely aside. Use the brand box to keep the motherboard on top of it and use it like a temporary station for installing all the components on it. The motherboard brand box is usually safe and the motherboard can be placed on it.

Don't place the motherboard on the metallic cover of the computer case or on any conductive surface. The motherboard comes with a battery and some of the components like the CMOS chip is always powered on. Placing the motherboard on metallic or conductive surfaces could cause a short-circuit because the contact pins of the motherboard components are on the back side.

Installing the Processor

The installation procedure of a processor can vary depending on the make of the processor and the socket type. Usually the processors for AM4 and LGA sockets have almost similar installation procedures. First, release the retention arm(s) of the socket by gently pulling it toward one side and then lift it all the way up. Also move the retention plate out of the way. Unbox the processor and hold the processor by touching only the edges, but never touch the contacts. Touching the contacts of the processor could cause moisture from the skin of the finger to interfere with the contacts, and could also cause corrosions in the future.

Align the golden triangle present on the processor with the triangle mark present on the socket of the motherboard. Sometimes notches are also present on both sides of the processor, which also needs to be lined up with the ones present on the socket. After aligning the processor with the socket, place it and let it fit on the socket by itself (don't apply any pressure). After placing it on the socket, gently tap on the processor to ensure it's perfectly in contact with the socket.

After successfully placing the processor, push the retention arm(s) all the way down and lock it (them) to the socket by pushing it toward the socket. This will finally release the socket cover and can be removed safely, making the top side of the processor exposed. If the computer needs an AIO cooling system, then the CPU fan that comes with the processor is not required to be installed.

To install the CPU fan for a computer without an AIO cooling system—first take the fan module that came with the processor and check whether it has plastic pins or metallic brackets (hooks). To install the CPU fan with four plastic pins—first rotate the pinheads (with arrow marks) in clockwise direction as much as possible. Then align the pins with the holes on the motherboard around the CPU socket and then place the fan on the motherboard and above the processor. After placing the fan on the motherboard, simply apply pressure on both corners of the CPU fan by pressing the two heads of the plastic pins at the same time. This needs to be done for the other two heads as well. Then finally check by holding the CPU fan and lifting the motherboard by it. Also look below the motherboard and check whether the plastic pins have gone through the holes properly.

In case there are hooks on both sides of the CPU fan, first place the CPU on top of the processor and hook the one side and then hook the other side. Then finally lock the fan by rotating the latch.

After installing the fan, connect the CPU fan wire with the connector pins on the motherboard. Look for the connector on the motherboard that says "CPU fan." Ensure that the wire doesn't touch the fan and always keep the wire somewhere away from the fan.

To install a processor for the TR4 socket like for the Threadripper—the process is somewhat different. A Threadripper processor comes with a screwdriver that needs to be used for screwing and unscrewing the screws on the TR4 socket of the motherboard. First, unscrew the three screws on the socket in the order mentioned by the manufacturer. The retention bracket will lift up by itself, and then lift up the lid below it and remove the transparent plastic cover from the metallic frame.

Then take the Threadripper processor but don't remove it from the plastic tray. Slide the plastic tray into the metallic frame of the socket. Then remove the CPU cover from the socket and lower the frame along with the processor (with the contacts facing down). Then lower the retention bracket all the way down. Finally secure the bracket by tightening the three screws in the reverse order one by one.

Most CPU fans come with a thermal paste that sticks to the processor after installation. However, some enthusiast-level processors don't come with a CPU fan with thermal paste on it. In such cases, a separate thermal paste should be applied on the processor before installing the CPU fan. The thermal paste sold in the market comes in a syringe and should be applied on the processor by creating two lines diagonally crossing each other forming a cross, and also put a pea size paste on the four sections between the lines

To clean up any unwanted thermal paste from the processor—a 99 percent pure isopropyl alcohol should only be used with a paper towel or microfiber. Never use any detergent or chemicals with tissue paper to clean the processor.

Every motherboard comes with a user manual that can be found inside the brand box or it can be downloaded from the manufacturer's website. More details about installing a processor can be found in the user manual of the motherboard.

Installing the RAM

In order to install the RAM modules, it's very important to check how many DIMM slots are present on the motherboard and whether they are single or dual-channel slots. This can be figured out by looking at the color codes of the memory slots. Most dual-channel slots have two different colors—two slots in one color (usually black) and the other two in another color (blue, red, green, etc.).

If the motherboard supports dual channel and has four DIMM slots with two different colors, then there are two options—either two RAM modules of same capacity can be installed in two different colored slots (to get the maximum bandwidth) or four RAM modules of same capacity can be installed in all the four slots.

Never mix RAM modules of different capacities and frequencies because that could give undesirable results and could even slow down the system and would force the system to run at single channel. However, a single RAM module of maximum (for only one slot) supported capacity can be used in one single slot. For example, if the motherboard supports 32 GB of maximum memory and has 4 DIMM slots, then only an 8 GB of RAM module can be installed in one slot.

To install the RAM module—first open the ejection levers present on both sides of the DIMM slot by pressing them down and out of the way. Then after unpacking, take the RAM module and hold it by touching the circuit board only (if it's not having any cover), but don't touch the contacts present at the bottom of the RAM module. Then align the notch of the RAM module with the plastic divider present on the DIMM slot. If it doesn't align then try again by flipping the RAM module. Once aligned, press it firmly down (on both top edges) until it fits perfectly in the slot and a clicking sound may be heard when the ejection levers on both sides will retain and lock the RAM module in place.

This process should be done for every RAM module that needs to be installed in the DIMM slots. To remove a RAM module from the slot for any reason, the ejection levers should be pushed all the way down, and the RAM module will be ejected.

INSTALLING THE MOTHERBOARD IN THE COMPUTER CASE (MARRIAGE)

To make the computer a complete one-piece machine—the motherboard should be installed inside the computer case. After both the case and the motherboard are ready, it's time to install the motherboard inside the case. This process can also be called "marriage," which is similar to that of car manufacturing. While manufacturing a car, when the chassis, transmission, and engine are inserted in the body, it's also called "marriage."

Before installing the motherboard, lay the case flat on the desk. First, take the standoff screws that came with the case and place them inside the case, and tighten them using a screwdriver. Then take the I/O shield that came with the motherboard and install it at the back side of the case. Next, install the motherboard by keeping it at an angle toward the I/O shield, so that the I/O ports get fit properly and place the rest of the board on the standoffs. Check whether all the standoffs are visible through the screw holes of the motherboard. When everything seems fine, use the screws to secure the motherboard and put them one by one over each standoff.

Finally, tighten them with a screwdriver, but be careful not to apply too much force while tightening them because while doing that sometimes the screwdriver might slip and hit the components of the motherboard—causing scratches or damage.

Once the motherboard is installed properly, all the data connections for the SATA drive(s) should be done with the motherboard using the SATA cables (often provided with the motherboard). Connect one end of the SATA cable to the SATA port of the drive and the other end to the SATA port on the motherboard. If the computer case has any USB and audio ports, then the wires from these ports should have a female header pin connector, which should be connected to the male header pins on the motherboard that says, "USB and audio."

Apart from that, connect the power on and reset (if present) buttons along with the LED indicator(s) to the motherboard system panel connector pins that have labels indicating power switch and LED as well. Refer to the motherboard's user manual for these connections.

After making all the data connections, make all the power connections. The PSU comes with a twenty-four-pin power connector—connect that with the twenty-four-pin connector present on the motherboard. The PSU also has an eight-pin connector for the CPU—connect it to the eight-pin connector present on the motherboard, and most ATX motherboards have this connector close to the CPU.

Connect the power connectors of the PSU to the drives mounted in the case. For a 3.5-inch drive, use the white connector with four wires (one red, two black, and one yellow), and use the black flat-type connector (either same colored wires or only in black) to connect a 2.5" drive.

Any cooling fan mounted in the case can either be connected to the motherboard (if there is any connector present on the motherboard) or it can be directly connected to any open female header connector of the PSU by cutting off the incompatible header of the cooling fan and connecting a male SATA power connector header to it.

INSTALLING THE AIO COOLER

Keep the AIO cooling system ready for installation. Most of the AIO coolers come with a set of parts that need to be assembled before installing the whole system. It comes with a step-by-step assembling guide and follows that guide first to assemble the AIO cooling system. The AIO cooling system mainly consists of a water pump/cold plate, connecting tubes, and a radiator. The cold plate will be mounted on top of the processor, which will allow cooling. The connecting pipes will circulate water to and from the radiator and through the pump, and the radiator will help exchange the heat.

After the assembling is complete, try to understand where and how the AIO radiator should be mounted. Find the ventilation panel either on top or front of the computer case. Mount the radiator with the screws provided and ensure that the direction of the air flow from the fans of the radiator is toward the interior of the case but not going outside because air outside the case is always cooler than inside. It's best to keep the fans in front of the radiator instead of keeping them back because that will help in better dust management.

Mounting the radiator at the front of the case is the best position because the fans will be able to cool the radiator better in that position than mounting it at the top. If the radiator is mounted at the top of the case, the hot air from the CPU or the graphics card would rise up and stay close to the radiator, which will not allow effective cooling.

After selecting the suitable place inside the case, mount the AIO radiator with the screws and keep it ready to be connected with the motherboard and the CPU. Organize the wires of the PSU and the AIO properly and thread them through the sides if necessary.

Take rubbing alcohol (greater than or equal to 99 percent isopropyl alcohol) and apply it on a paper towel. Then clean both the surfaces of the processor (top exposed part only) and the cold plate with the paper towel. After cleaning the surfaces and making them free from any dust or debris—apply the thermal paste on the surface of the processor creating two lines diagonally crossing each other forming a cross, and also put a pea size paste on the four sections between the lines. Then place the cold plate on top of the processor and either secure it using the screws or hook it using the hooks (whichever installation mechanism is available as per the type of processor).

Lastly, connect the power connector of the pump and the RGB lighting (if present) to the motherboard.

It's always not necessary to use an AIO cooler and some CPU coolers can also better cool the processor. CPU coolers like the Noctua CPU Cooler (NH-U12S) can cool the processor better than an AIO cooler and also cause less operating noise.

The motherboard may come with an M.2 slot for installing an M.2 SSD and it should be installed as per the instructions mentioned in the user manual. Some motherboards also come with a Wi-Fi antenna that needs to be installed too.

INSTALLING THE GRAPHICS CARD

To install the graphics card—first, remove the screw of the bracket cover of the computer case next to the PCIe slot (x8 or x16) of the motherboard, where the graphics card will be installed. After removing the bracket cover, grab and hold the graphics card by the cooler (without touching the contacts or the electronic components) and line up with the slot, keeping the video port of the graphics card toward the expansion slot (the one without the cover). Then push down each end of the graphics card into the PCIe slot and a click sound might be heard when the graphics card gets locked in place by the retention clip.

To remove the graphics card for any reason, push the clip or release the clip sideways (depending on the type of clip present on the motherboard). Finally, put the screw back to the same location and tighten it over the metallic bracket of the graphics card so that it gets attached to the computer case properly.

To install more than one graphics card on more PCIe slots, the same installation process can be repeated. Two to four graphics cards can be connected using the AMD crossfire (for AMD graphics cards) or Scalable Link Interface (SLI) technology (for Nvidia graphics cards).

A graphics card can either have a six-pin or twelve-pin power connector that needs to be connected with the power cables of the PSU.

After making all the wired connections, use zip ties to tie any freely hanging or sagging wires together and organize them in such a way that there shouldn't be any problem with air flow inside the case. Tie the same type of wires and group them together, so that during any maintenance or future upgrades, it shouldn't be a problem finding which wires go where. After tying the wires, cut the excess part of the zip ties using the wire cutter to make it look nice and tidy.

POWERING ON THE COMPUTER AND BIOS SETUP

After assembling the computer, connect the keyboard and mouse to the computer using the USB ports. Connect the speakers (if present) with the audio out connector of the computer. Connect the display to the computer using the VGA or HDMI cable. Lastly, connect the AC power cable to the PSU and plug it to the AC outlet. Then check whether there is any switch on the back side of the PSU—if it's there, then turn it on.

When ready, press the power button of the computer. It might take a few seconds and if everything is fine, it might give a single beep (may not be heard on some motherboards and depends on the make). Finally it will show a splash screen (usually the logo of the manufacturer), which indicates that the computer build was successful.

In case there is no display on the screen or any beeping sound is heard, then refer to the BIOS beep codes and troubleshooting steps mentioned in the user manual.

Even if the computer powers on successfully with a splash screen, at this point the computer can't be used for any work. In order to make a computer work with apps, any OS like Windows or Linux needs to be installed.

The computer can be configured using the BIOS setup, and it can be accessed by tapping a particular function key of the keyboard right after the computer is powered on, and the function key number for BIOS setup may be displayed with the splash screen or it can be found in the user manual.

Many types of hardware settings like boot order, HT, CPU overclocking, and so on can be configured using the BIOS setup.

Apart from a desktop computer, a single-board computer like Raspberry Pi can also be built easily without the need of too many hardware parts, but it can't be used like a high-end computer and neither has any option to install a graphics card or upgrade the hardware. That's why such a computer can only be used for engineering projects and experiments.

Chapter 14

How to Install an OS and Apps on a Computer

After building a computer, it will need an OS and some apps to make it usable. Without the OS and apps, a computer can't be used for any work. However, a pre-built computer bought from a vendor could come with a pre-installed OS and apps. But for any reason if that needs to be reinstalled or changed to another OS, then a fresh copy of the OS should be installed.

The installation process of almost any OS is somewhat similar and can be done using the image file of the OS or any installation disk bought from the market. In case the OS is downloaded from the internet, the OS can be installed using an installation media like an optical disk or USB flash drive. Nowadays most computers don't come with an optical disk drive and using a flash drive to install the OS is the only option.

The OS installation process using an USB flash drive basically requires two things—the OS image file (ISO) and a bootable USB flash drive creator tool like Rufus. This also requires another computer with internet access, through which the image file can be downloaded and also a USB flash drive that can be used as the installation media. But ensure that the capacity of the flash drive is more than the size of the image file and should have a good transfer speed. That's why a flash drive with 8 GB of class 10 is ideal for this purpose.

If a CD/DVD needs to be used as an installation media, then the computer that will be used for creating the installation media must have a writable DVD drive with a burning app like Nero Burning ROM installed.

OS INSTALLATION PROCESS

The installation process of Windows and Linux can be possible on an x86-64 computer with Intel or AMD processor using a bootable installation media. But the macOS can't be installed on such a computer in the same way. The installation of Windows and Linux (like Ubuntu and Pop!) can be done using a few basic installation steps with the help of some software tools and methods. However, before installing any OS, the recommended system requirements should be checked.

The installation process includes downloading the image file, creating the installation media, booting the computer from installation media, installing the OS, and thereafter installing the drivers, updating the OS, and optimizing the OS (see figures 14.1 and 14.2).

Downloading the Image File

After deciding which OS needs to be installed, the next step is to find the image file (usually an ISO file) for it. To find the download link for the image file, search in Google using the name of the OS and then go to the developer's webpage, to download the image file. The image file of Windows OS can be downloaded from Microsoft's website, but for Linux, there are many different websites for various Linux distributions (distros).

The image file of Linux may be available for many types of system architectures like ppc64le, s390x, ARM64, and x86-64. Choose the one that is compatible with the computer—most desktop computers comply with x86-64 system architecture.

Creating the Installation Media

After downloading the image file, it needs to be either burned to an optical disk (if installing from an optical drive) or written to an USB flash drive. The image file like an ISO file can be burned to an optical disk like a CD/DVD using a burning app like Nero Burning ROM. But if the image file needs to be written to an USB flash drive then a bootable USB creator app like Rufus needs to be used.

Rufus can be downloaded from the developer's website that can be found by searching on Google. After downloading the app, open it to start with the image-writing process. Before writing the image file on the USB flash drive, ensure it is formatted to the FAT32 file system. If not, then format it. Then on the Rufus, select the image file by clicking the "SELECT" button. After selecting the ISO file, ensure that the name of the USB flash drive (along with the capacity) is selected and appears under the "device" option. Under

How to Install an OS and Apps on a Computer   123

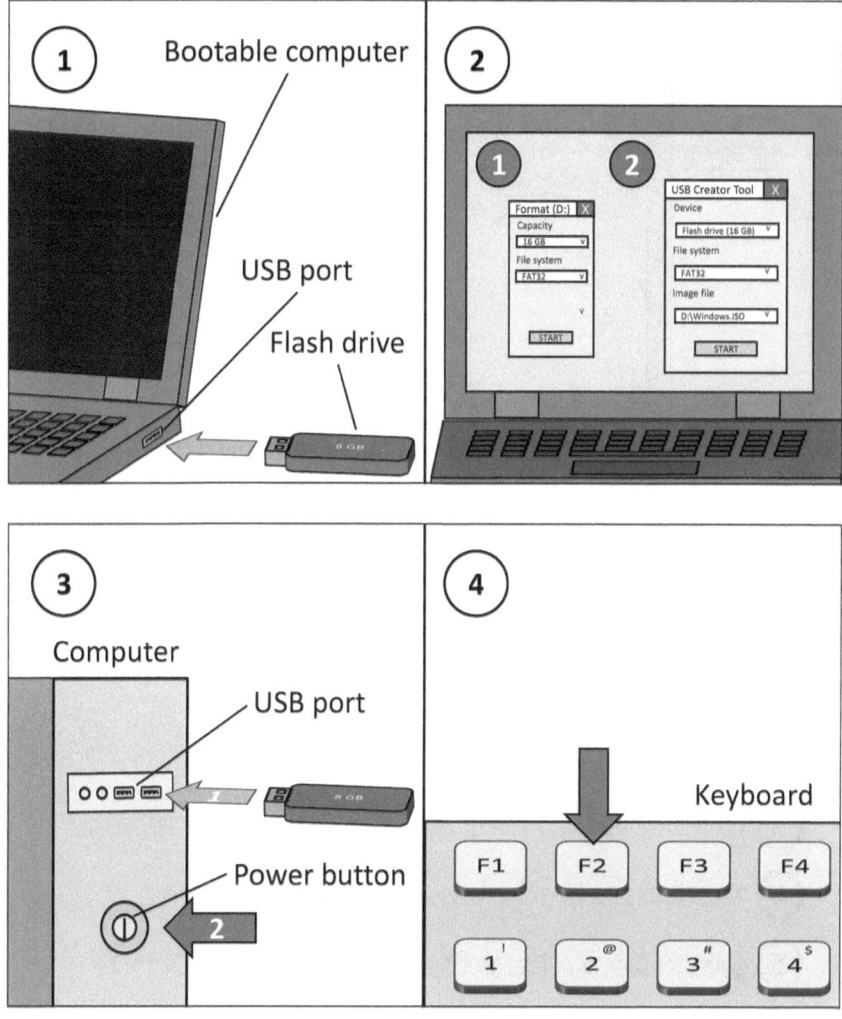

Figure 14.1 Creating a Bootable USB (Part 1). Author Created.

the "boot selection" menu, select the "disk or ISO file" option. Keep the file system selected as FAT32 only.

After setting up everything, click the "START" button to start the writing process, which could take some time depending on the size of the image file. After the writing process gets done successfully, a message will appear. Close the Rufus app and check the flash drive by opening the drive and it should contain all the installation files. Then safely remove the USB flash drive (when used on a Windows computer) by right clicking on the drive letter and choosing the eject option, which can also be done from the system tray.

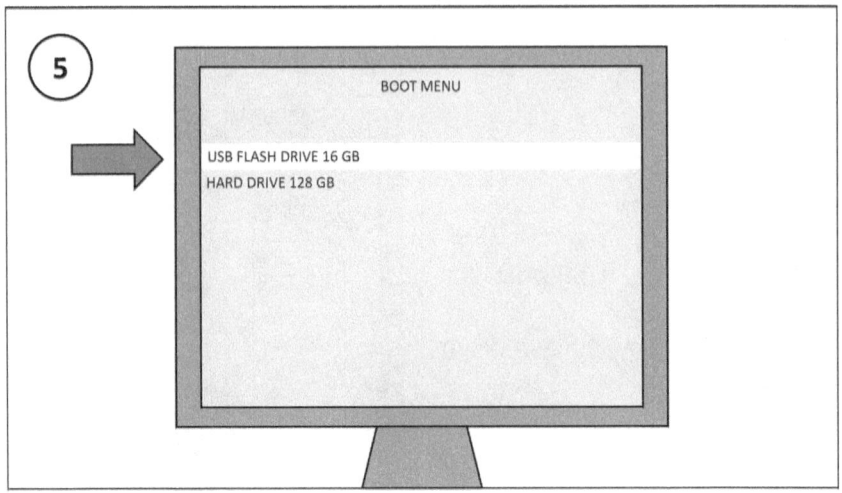

Bootable USB Creation and Booting Process:

① Connect the USB flash drive to a bootable computer.

② Format the flash drive and write the bootable OS image.

③ Connect the USB flash drive to the computer and press the power button.

④ Keep tapping the function key (usually F2) to select the boot menu.

⑤ Select the flash drive from boot menu and press enter.

Figure 14.2   Creating a Bootable USB (Part 2). Author Created.

Booting from Installation Media

To install the OS from the optical disk or USB flash drive, the computer needs to be booted with that bootable media. This can be done either by selecting the boot menu during startup or the first boot device can be changed in the BIOS settings. To access the BIOS setup, the specific function key (usually F2) should be tapped repeatedly right after pressing the power button (to power on the computer). But don't keep pressing the power button—press it only once.

To know which function key can access the BIOS setup, refer to the user manual of the pre-built computer or the motherboard (in case of a custom-built computer). After accessing the BIOS setup, change the boot order and keep the optical drive or the USB device as the first boot device, and restart the computer after saving the setting by pressing the function key (usually F10) that saves the BIOS settings. Then the computer will boot from the bootable media automatically.

Another way to boot from the media is to select the boot menu by tapping the function key (usually Esc/F10/F12) that brings up the boot menu but refer to the user manual to know the exact key. This is the simplest way to boot any USB device, and can be done without changing the BIOS settings. When the boot menu appears, select the optical disk or the USB flash drive from the menu to boot from that media.

Installing the OS

After selecting the installation media from the boot menu, the computer will boot from the installation media. In Ubuntu, a different selection option might appear—select "Try or Install Ubuntu" from the menu. A splash screen will appear during the booting process, and after sometime it will show a GUI environment with an installation window.

In Windows, the splash screen will appear while booting from the media. Depending on the type of OS, the installation options may vary. In Ubuntu, there will be an option to select whether to try or install. Select the preferred language and click Install. In both Ubuntu and Windows, there will be options like language and keyboard layout that need to be selected as per the requirements. In Windows, the product key should be entered to get all the features, but if it was not purchased, then the trial version can be installed without the key.

If the same version of Windows that came with a pre-built computer is going to get installed again, then without using the product key it can be installed. Next, a user license agreement will appear, which needs to be agreed.

After going through the installation process, it will ask to select the drive where the OS needs to be installed. It's better to format the drive and select the drive to install the OS. In Ubuntu, the option to erase the drive will appear in the "Installation type" window of the installation process.

In Windows, after completing the installation, the system will restart automatically and post-installation setup windows will appear one after another, asking for details about the user and the computer, and depending on the version of Windows, the setup requirements may be different. Refer to Microsoft's website or search on Google about setting up the version of Windows before installing to know exactly what to configure. During this setup, it will ask a few things like device name, username, password, and so on. After completing all the setup process, Windows will login to the user account and finally the desktop will appear.

In Ubuntu, all the user details need to be entered during the installation process and after installation it will finally boot to the desktop.

Installing the Drivers

After installing the OS, the computer might need the drivers to function properly. In case of any recent version of Windows, the drivers for audio and video might get installed automatically after installing the OS. But it's better to install the drivers that came with the hardware because the drivers provided by the manufacturer will have better compatibility. The drivers can be found in the optical disk that came with the computer or the hardware, and can also be downloaded from the manufacturer's website.

In the case of Linux, the condition can vary depending on the distro. When Ubuntu is installed, almost all the drivers for the hardware, including the display, audio, and Wi-Fi, get installed automatically. However, some Nvidia graphics cards don't have an open-source driver for Linux. That's why any other compatible Nvidia driver needs to be installed, which can be done using the driver manager. It's also advisable to update all the Linux drivers after installing Linux.

Updating the OS

After installing the OS, it's important to update the OS and it can be done either manually or automatically. The manual process of updating Windows can be done by selecting Start > Settings > Windows Update. But the automatic update can work only during the selected schedule. The update process may take from a few minutes to several hours depending on the size of the files being downloaded during the update process, and can also vary depending on the speed of the internet.

When the installed OS is Ubuntu, the update can be done by using the Software Updater or by using the commands "sudo apt update" and "sudo apt upgrade" at the terminal.

Optimizing the OS

After installing and updating the OS, it's time to optimize it for the best performance. This needs to be done especially for Windows, because this OS behaves differently on computers with different hardware configuration. If Windows was installed on a low-end computer, then a few changes need to be made to the OS. First, the OS should be optimized for best performance, and to do that go to System Properties > Advanced > Performance Options. From there select "Adjust for best performance" and click Apply and then click OK. Click OK again to exit the System Properties.

After completing any OS update, unnecessary files might be stored on the hard drive and this could slow down the system if there is not enough space or if the primary hard drive is a magnetic one. Most magnetic drives suffer from fragmentation after large files are stored. That's why a disk cleanup and defragmentation are necessary.

To clean the drive in Windows, type "cleanmgr" in the Run box and click OK or type it in the search bar and select the disk cleaning utility. Choose the primary root drive (usually C: drive) and click OK. After analyzing the drive, it will show a list of files that can be cleaned. Don't clean these, but click the "Clean up system files" button.

Then again it will analyze the drive and finally it will show a list of files (including the system files) that can be cleaned. Click OK to start the disk cleaning process and it might take some time depending on the size of the files. Once cleaned, the disk cleaning utility will exit by itself.

To clean the temporary files on Ubuntu, use the commands "sudo apt-get clean" and "sudo apt-get autoremove" at the terminal.

Besides disk cleanup, it's also important to defragment the drive when the drive is a magnetic one. But defragmentation won't help improve the performance of any SSD drive. It can be done either by using the defragmentation utility of Windows or using a third-party defragmenter like PerfectDisk. To use the defragmentation utility of Windows, type "dfrgui" in the Run box and click OK or type it in the search bar and select the utility. Choose the primary root drive and click Optimize to start the defragmentation process. This defragmentation process can also be scheduled to get done automatically on a regular basis.

Defragmentation is also possible in Linux but it's not necessary because the file system is superior to NTFS or FAT32 and doesn't easily get affected from fragmentation.

APPS TO INSTALL

After installing the OS, the computer becomes ready to be used for any work. But to use a computer for a specific purpose, an app needs to be installed too. Apps for Windows can be installed either by downloading from the internet or can be installed from the Microsoft app store—either free or paid.

A computer can either be used for everyday work or for any business or educational purpose. That's why it's the apps that will help perform specific tasks as per the requirement. But whatever may be the purpose of use, a few bunch of apps will always be required and must be installed. Among these apps include the security app (like an antivirus), utility apps, multimedia apps, communication apps, and office apps. Some of these apps may come with the OS for free but there are better apps that can be used additionally.

- Security apps—this type of apps can protect the computer from malware and hacking attacks. Antivirus apps like Avast need to be installed to keep the computer secure. Linux is less prone to malware and keeping an antivirus running on Linux could consume RAM space and cause CPU usage unnecessarily.
- Utility apps—with the help of utility apps many types of accessibility and maintenance works can be performed. Apps like disk cleaners, defragmenters, registry cleaners, app uninstallers (like Revo uninstallers), and so on, are needed to do regular maintenance. Registry cleaners like RegCleaner should be used after uninstalling any application, although Revo Uninstaller can also do the job.

  WinZip is an app that is required to access the zip files and can also be used to pack too many files into a single zip file. WinZip is not a freeware and can be used as a trialware but 7zip is a freeware that can be used as an alternative.

  NetBalancer is another useful app that can be used to monitor and control network traffic. This app comes with a bandwidth toolbar that appears on the taskbar and can be very useful while downloading or uploading files. It can also let the user know about the internet speed while using the internet.

  A backup app like EaseUS Todo can also be installed to backup files regularly. Although Windows come with a backup utility, using a third-party backup app could provide additional features.
- Multimedia apps—such apps are required to process various types of data containing text, images, audio, and video. Media players like the media player classic (MPC) and AIMP player can be installed to play various formats of audio and video files. To view image files, the default app that comes with Windows, called Microsoft Photos can be used, but Windows Image Viewer (Picture and Fax Viewer) available in previous versions of Windows was better and light. Pineapple Pictures is one such lightweight portable image viewer that can be downloaded from GitHub.

  Such multimedia apps also come with Ubuntu and more apps can be also installed as per the requirement.
- Communication apps—this type of apps can be used to communicate over a network or the internet. VoIP apps like Skype are used to make internet calls and such apps are often required to communicate with others. Apart from that, email client apps like Microsoft Outlook and Mozilla Thunderbird can be used to send and receive emails. Download managers are also

required to download various types of files. uGet is one good download manager that can be downloaded for free from GitHub.

Web browser is also one multimedia app that can display web pages with text, images, audio, and video. Windows usually came with Internet Explorer (IE) as a default web browser but now Microsoft Edge is the browser that comes with Windows 11 along with IE. Other browsers like Google Chrome and Mozilla Firefox can also be installed and are more compatible with many types of websites.

- Office apps—such apps are required for creating and viewing documents. Microsoft Office is the most popular office suite that everyone likes to use and it comes with a bundle of apps like Word for creating drafts, Excel for creating spreadsheets, PowerPoint for creating slides, and so on. However, this suite can only come with a pre-built computer with a license but it's not free. LibreOffice is a similar office suite that can be downloaded for free and can also be used on both Windows and Linux.

A PDF viewer is also necessary to view PDF files, and Adobe Acrobat Reader DC can be downloaded for free to view any PDF file.

After installing the apps, it's also important to disable some unnecessary apps to run in the background. To do that, go to Settings > Apps > Startup. Then select the app that is not required to run at the startup and select Disable. Ensure that any essential app that is required to run in the background like an antivirus should not be disabled.

It's better to install the antivirus software before downloading or installing any apps, because most freeware and pirated apps could come with malware. However, this problem may not always happen with a Linux OS like Ubuntu, but keeping the antivirus active on Linux can protect the system not only from any malware but also from hacking attacks.

Chapter 15

How to Use a Computer for Homework and Teaching

Teachers and students can use a computer for education, and the good news is that computers for education can be quite affordable nowadays. Thanks to the cutting-edge technologies used for manufacturing computer hardware, people can now buy or build a computer for everyday use within the budget. That's not all; nowadays the availability of broadband internet services like Google Fiber has facilitated computer users to use the internet for both business and education.

With the rise of Web 2.0 and various cloud computing services, the internet has become a great place for students and teachers. Students can learn about various subjects and do their homework using a computer and the internet. Teachers can also use a computer to create educational content for teaching and can attend online classes through the internet. This process of providing education over the internet not only makes learning convenient but also bridges the gap between students and teachers located at far-off places.

With the help of a computer and internet a student can get the best resources and tools from around the world to learn and study online. Teachers can also contact other teachers and researchers to gather more knowledge about a topic using the internet. Online classes can help students to learn and teachers to teach from any convenient location, and can also provide education during any pandemic.

Since the last few years the percentage of students taking online education has increased significantly and this has happened because of the arrival of many virtual schools like the Florida Virtual School (FLVS). Virtual schools like FLVS provide free online education to Florida students in kindergarten through twelfth grade. FLVS was founded in 1997 as the first statewide Internet-based public high school in the United States, but in 2000, FLVS was established as an independent educational entity by the Florida Legislature.

Even students who are enrolled in offline courses can take advantage of a computer and the internet for their studies. There are many types of educational apps that can be used to learn things and do homework. Such apps can be installed on a computer and can run independently, but some apps that are provided by cloud service providers require an internet connection. Apart from apps, students can use the internet to study and learn about their subjects. Nowadays the Google search engine has become a magical tutor that can help find articles and news containing information about various topics.

Teachers can also use various apps on a computer to do research and also for processing student's data. Even some cloud-based apps can be used for teaching purposes and can create a virtual classroom too. Google can also help teachers find new ways of teaching and can let them get information about various education online. The internet can even help teachers to search for jobs and get more employment opportunities (see figure 15.1).

Therefore, computers have evolved the traditional education system in such a way that it can be reached by anyone from anywhere using the internet.

APPS AND THE INTERNET FOR HOMEWORK

There are many kinds of apps that can be installed on a computer for doing homework. And for a student one single app may not help to learn about a particular subject while doing the homework. It will also depend whether the student is in school or college. For school students various apps can be used

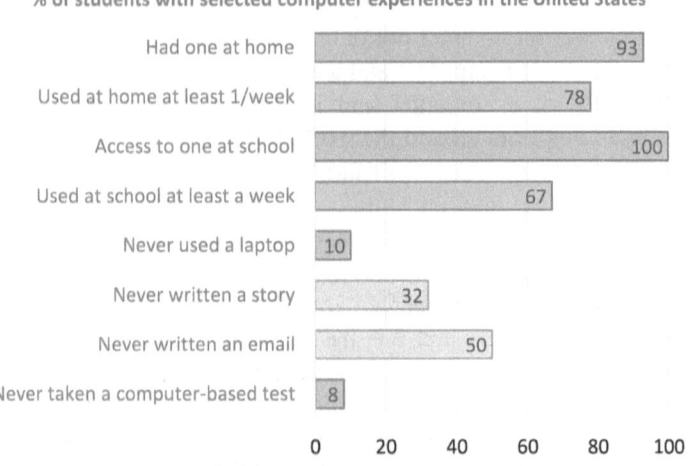

Figure 15.1 Percentage of Students with Selected Computer Experiences. Author Created.

for doing homework. But all apps may not run on a low-end computer. That's why ensure that the apps match the system requirements.

Depending on the subjects and type of work, an app can be used. For example, apps like GeoEnZo and GeoGebra can be used for doing homework related to geometry. On the other hand, the Virtual Chemistry Lab app can be installed from Microsoft's app store, which will help students learn about chemistry. That's why to find the suitable app for the subject it's better to search on Google.

Students can also use office suites like Microsoft Office or LibreOffice to create any documents for assignments or projects. Such office suites come with a word processor, spreadsheet, and a slide-show presentation program that can be used for creating documents with pictures, charts, drawings, and so on. Such documents can be converted to PDF files and can either be printed on a paper or shared online.

Any homework that requires drawing and artwork can be done using a vector editing app like Inkscape. Inkscape can be used to create both 2D and 3D images, and it can render any created 3D object with transformations like moving, rotating, scaling, and skewing. It also comes with many types of painting brushes and colors to choose from. Inkscape supports mathematical diagramming and comes with a computational geometry library called 2Geom.

Students who want to learn programming can use a computer for programming too. BASIC is the most common programming language taught in schools and any low-end computer capable of running Windows can be used to learn BASIC. The IDE along with the compiler can be downloaded with Microsoft's QBASIC from the internet. BASIC programming can also be used to solve many math problems and can be also used for doing homework too.

Students studying music can use a scorewriter app like MuseScore to create or edit sheet music (using musical notations), which can even be printed. LenMus Phonascus is another great app that can be used to study music theory, and it can be used to focus on specific skills and exercises, on both theory and aural training. Besides these apps, Rocksmith from Ubisoft is a music video game that students might love to play on a computer to learn about playing a guitar.

Apps like Google Earth can be used by students to study about the earth's geography and it comes with many features like satellite/map view, zooming, street/flight, and so on, can be used to explore the earth in various ways. Students can even get details about many historical places and view them in 360 degrees just like in real life without visiting them physically. They can even take pictures of any newly visited places and pin them on Google Earth.

Cloud computing services also provide various apps that can be used by students to do their homework too. Google translate is one such app that can be used to translate text between many different languages, and students can learn languages using this online tool. Evernote is another app that can be used by students for taking notes and creating schedules for their studies, which might be also helpful when the homework needs to be completed in time. Apart from these there are many such web apps that can be used by students for doing homework.

It also may become necessary for students to discuss their homework with their friends and teachers. This can be done using a videotelephony app (with whiteboard) like Zoom. There are many VoIP apps that can be used for video conferencing but the ones with a whiteboard are useful to draw and discuss about a subject matter. However, such apps require a broadband internet connection with good speed.

A good web browser like Google Chrome can be enough to let students study online. With the help of a browser students can access educational websites like Wikipedia to read about various articles related to their education. The Web 2.0 has made it possible to utilize the internet in multiple ways, and now it has become possible to do things like social networking, blogging, and vlogging.

Sometimes books are also required for doing homework and the internet can also be a great place to search for books. Various types of textbooks can be found at open-content sites like Open Textbook Library and Wikibooks. Books from such websites can be downloaded as a PDF file and can be read using a PDF reader app like Adobe Acrobat Reader.

If a student is not comfortable with using the soft copy of the book, it can be either printed using a printer or an original hard copy of the book can be purchased from Amazon. Amazon also provides an option to buy used books, which could be a good option for those students who don't have a budget to afford new books.

Students can be in touch with their friends and teachers using social networking sites, and can share tips and updates about their education. They can even visit online forums to discuss any problem related to homework and some community forums can be useful for that purpose. It's also possible to create blogs and share knowledge using various blog sites like Blogger.

Writing blogs can help improve writing skills, and sharing any writing work with friends and people around the world can make a student not only a better performer but also a good writer.

To store any writing work, students can use cloud storage like Google Drive, which can also be used to share any stored file with others by creating an URL link to the file.

There are many websites that run writing contests that a student can join to compete with other writers. However, this should not be the primary focus of a student but to be done along with the curriculum studies only.

Many online video sharing sites like YouTube can be used to watch educational videos, and can be a great place for learning. Students can even vlog to share about their life and experiences that could let other students and teachers know about it. There are many tutorial videos available on YouTube, and students can watch them to learn and can also ask questions as comments to get answers from others. This process of learning can improve the knowledge of the students only by using a computer with an internet connection.

APPS AND THE INTERNET FOR TEACHING

Teachers can use a computer for teaching using various types of apps, and most of them can be downloaded from the internet for free. But the selection of apps will depend on the type of teaching work and the subject of the education. Teachers may require many types of apps if they are teaching students of too many different courses. However, some online services like Google for Education provide all the apps that a teacher can use for education.

A teacher can also use the apps (that a student needs for doing homework) for preparing teaching content, but may require a few additional apps too. MathType is one such app that can be used by a math teacher to prepare any document containing mathematical symbols and expressions. There are many useful apps like SageMath that can be used by teachers to understand elemental math and can teach the students more about them.

Many simulation apps like Project Chrono and Open Source Physics (OSP) can be used to simulate physics and teachers can demonstrate various physics simulations to students. ACD/ChemSketch is another app that can be used to teach about chemical formulas and molecular bonding, and this app can be used to create and modify images of chemical structures. There are also many planetarium apps like Stellarium and World Wide Telescope (WWT) that can be used to explore the solar system and some parts of the universe virtually, which can be used by the teachers to teach about the universe.

Teachers would also require an office suite like Microsoft Office or LibreOffice to create any documents for teaching or educational purposes. They can not only use the word processor for creating drafts but also can use the Excel program to keep records of students and their scores. The PowerPoint presentation program can be used to create slides for teaching in the classroom or can even be shared online with others. The Learning Essentials for Microsoft Office On-Demand is an add-on feature that can be installed along with the Microsoft Office 365 to get more features of education.

There are also many apps that are used for productivity and may also be used for teaching. Apps used for video and photo editing can also help teachers to create content for students. Apps like Adobe Photoshop and CorelDRAW can be used by teachers to edit and create content for students to ignite their imagination and make their learning easier. CorelDRAW Graphics Suite is one award-winning app suite that can be used by teachers to create educational content for teaching or publication purposes.

Video editing apps like Adobe Premiere Pro and Vegas Pro can be used for creating video content for education that can be used for teaching either online or offline. Teachers can use a CAD app like Autodesk Fusion 360 to create 3D models for engineering and research. Audio editing apps like Audacity and Sound Forge can be used to create any audio content for education. However, most of these apps will require a good hardware configuration and a low-end computer may not be suitable to run them smoothly.

The best way to use a vector graphics editor like Adobe Illustrator or CorelDRAW is to use a touch screen or a graphics tablet with the computer, which will allow a stylus pen to draw and use the apps.

Music teachers can use scorewriter apps like MuseScore to create or edit sheet music (using musical notations), which can even be printed. There are other professional scorewriter apps like Dorico from Steinberg that can be also used for professional music creation. Besides a scorewriter, there are apps like Magix Music Maker and FL Studio that can be used to create music on a computer without even having a physical musical instrument. However, some professional music composers use a MIDI keyboard with a computer to create music along with such apps.

Communication apps like Skype and Zoom are also essentials for the teachers, and as the Zoom app comes with a whiteboard feature, it can be used to teach students while taking online classes too. MURAL is one special whiteboard app that can be used for education, and teachers can make use of the app to help the students learn while taking any online classes. Besides these apps, there are also many cloud apps that can be used by the teachers for teaching purposes. Such apps can be searched on Google and should be selected as per the requirements.

Teachers can use Optical Character Reader (OCR) and Optical Mark Reader (OMR) apps to process educational documents like exam papers and handwritten notes. OCR apps like GOCR can convert any image files to text and can be used to store scanned documents as text. OMR apps like FormScanner can be used to process forms and exam papers of multiple choice types. Such OCR and OMR apps can make a teacher's work easy, and some apps can even automate the checking processes of too many documents. That's why in some schools and colleges such apps are custom built to meet their processing requirements.

Web 2.0 and cloud computing services have also facilitated the teachers to use the internet for teaching. There are many web apps that can be used for teaching purposes also. To make use of the internet, teachers just need to use a web browser like Google Chrome or Mozilla Firefox. Apps like Jamboard can be used for teaching, and it's a digital interactive whiteboard that the teachers can use to teach the students using the Google Workspace. Apart from a whiteboard app, a project management app like HeySpace from TimeCamp Inc. can be used to collaborate students to manage their projects and assignments.

Teachers can also take advantage of blog sites like Blogger to post educational content for students and enthusiasts. Vlogging is another great way to share video content using video-sharing sites like YouTube, and teachers can create video lectures and educational content (using video editing apps) that can be shared on such video-sharing sites, which could be helpful to the students.

Cloud storage like Google Drive can also be used by teachers for storing and sharing files online with students, but can also be used to backup any content with large file sizes like videos and apps. However, the storage space of most free file-hosting services may not be sufficient enough to backup large files regularly and may require buying more storage space in the future.

There are many online websites that offer online teaching jobs that a teacher can join to earn from teaching. Companies like Tutors.com that provide online courses also give opportunities to teachers to earn from teaching online.

A computer is a digital machine that can be used for doing various types of work, and it's the perfect combination of software and hardware that makes a computer useful for a specific work. A computer is used for various applications like entertainment, business, medical, and so on but it can also be used for education. Using a computer effectively can help improve the knowledge of teachers and students brilliantly.

Glossary

μm—Micrometer

2D—Two-Dimensional
3D—Three-Dimensional
5G—Fifth Generation

ABC—Atanasoff–Berry Computer
AC—Alternating Current
ADSL—Asymmetric Digital Subscriber Line
AI—Artificial Intelligence
AIM—AOL Instant Messenger
AIMP—Artem Izmaylov Media Player
AIO—All In One
ALU—Arithmetic Logic Unit
AMD—Advanced Micro Devices
AMOLED—Active-Matrix Organic Light-Emitting Diode
AOL—America Online
APFS—Apple File System
API—Application Programming Interface
APU—Accelerated Processing Unit
ARM—Advanced RISC Machine
ARPANET—Advanced Research Projects Agency Network
AT&T—American Telephone and Telegraph Company
ATA—Advanced Technology Attachment
ATIC—Alphanumeric Television Interface Controller
ATX—Advanced Technology eXtended

BASIC—Beginners' All-purpose Symbolic Instruction Code
BC—Before Christ
BCI—Brain Computer Interface
BGA—Ball Grid Array

BIOS—Basic Input/Output System
BNC—Bayonet Neill–Concelman
BSD—Berkeley Software Distribution

CAD—Computer Aided Design
CD—Compact Disc
CIL—Common Intermediate Language
CISC—Complex Instruction Set Computer
CLI—Common Language Infrastructure
CLR—Common Language Runtime
CMOS—Complementary Metal-Oxide-Semiconductor
CP/M—Control Program/Monitor
CPU—Central Processing Unit
CRT—Cathode-Ray Tube
CU—Control Unit
CUDA—Compute Unified Device Architecture

DCCP—Datagram Congestion Control Protocol
DDR—Double Data Rate
DEC—Digital Equipment Corporation
DFD—Data-Flow Diagrams
DHCP—Dynamic Host Control Protocol
DIMM—Dual In-line Memory Module
DMI—Direct Media Interface
DNS—Domain Name System
DOS—Disk Operating System
DPI—Dots per Inch
DRAM—Dynamic Random-Access Memory
DVD—Digital Video Disc

eDRAM—Embedded Dynamic Random-Access Memory
EEPROM—Electrically Erasable Read-Only Memory
EFLOPS—ExaFLOPS
ENIAC—Electronic Numerical Integrator and Computer
EPIC—Explicitly Parallel Instruction Computing
ESD—Electrostatic Discharge
EUI-64—Extended Unique Identifiers
EV—Electric Vehicle

FAT—File Allocation Table
FHD—Full High Definition
FLOPS—Floating-Point Operations Per Second
FLVS—Florida Virtual School
FPU—Floating-Point Unit
FSB—Front-Side Bus
FTP—File Transfer Protocol

Glossary

GB—Giganyte
Gbit/s—Gigabits per second
GDDR—Graphics Double Data Rate
GEOS—Graphic Environment Operating System
GHz—Gigahertz
GM-NAA—General Motors and North American Aviation
GPU—Graphics Processing Unit
GTP—GUID Partition Table
GTX—Giga Texel Shader eXtreme
GUI—Graphical User Interface

HD—High Definition
HDD—Hard Disk Drive
HDMI—High-Definition Multimedia Interface
HP—Hewlett-Packard
HT—Hyper-Threading
HTML—Hypertext Markup Language
HTTP—Hypertext Transfer Protocol
HTTPS—Hypertext Transfer Protocol Secure

I/O—Input/Output
IA-64—Intel Itanium Architecture
IBM—International Business Machines
IC—Integrated Circuit
ICMP—Internet Control Message Protocol
IDE—Integrated Development Environment
IEEE—Institute of Electrical and Electronics Engineers
IGMP—Internet Group Management Protocol
IGP—Integrated Graphics Processor
IMAP—Internet Message Access Protocol
Intel Xe—Intel eXascale
iOS—iPhone Operating System
IP—Internet Protocol
IPC—Inter-Process Communication
IPS—In-Plane Switching
ISO—International Organization for Standardization
ISP—Internet Service Provider
ITO—Indium Tin Oxide
ITX—Information Technology Extended

KB—Kilobytes
Kbit/s—Kilobits per second
kHz—Kilohertz

LAN—Local Area Network
LCD—Liquid Crystal Display

LED—Light-Emitting Diode
LEO—Low Earth Orbit
LGA—Land Grid Array
LINC—Laboratory INstrument Computer
LTE—Long Term Evolution

MAC—Media Access Control
MacOS—Macintosh Operating System
MAN—Metropolitan Area Network
MB—Megabyte
Mbit/s—Megabits per Second
MCI—Microwave Communications, Inc
MCST—Moscow Center of SPARC Technologies
MHz—Megahertz
MIDI—Musical Instrument Digital Interface
MIME—Multipurpose Internet Mail Extensions
MIPS—Million Instructions per Second
MISC—Minimal Instruction Set Computer
MITM—Man-In-The-Middle
MOS RAM—Metal-Oxide-Semiconductor RAM
MT/s—MegaTransfer per second

NAND—Not-AND
NAS—Network-Attached Storage
NDK—Native Development Kit
NFC—Near-Field Communication
NIC—Network Interfaces Controller
Nm—Nanometer
NOR—Not-OR
NSFNET—National Foundation Network
NTFS—New Technology File System
NVMe—Non-Volatile Memory Express

OCR—Optical Character Reader
OMR—Optical Mark Reader
OOP—Object-Oriented Programming
OpenCL—Open Computing Language
OpenGL—Open Graphics Library
OS—Operating System
OSI—Open Systems Interconnection
OSP—Open Source Physics
OSPF—Open Shortest Path First
OTP—One Time Programmable

P2P—Peer-to-Peer
PATA—Parallel ATA

PCIe—Peripheral Component Interconnect Express
PDP—Programmed Data Processor
PGA—Pin Grid Array
PnP—Plug and Play
PON—Passive Optical Network
PoP—Point of Presence
POP—Post Office Protocol
POST—Power-On Self-Test
PPP—Point-to-Point Protocol
PROM—Programmable Read-Only Memory (PROM)
PS/2—Personal System/2
PSU—Power Supply Unit

QBASIC—Quick BASIC
QDOS—Quick and Dirty Operating System
QPI—QuickPath Interconnect

RDMS—Relational Database Systems
RGB—Red Green Blue
RIMM—Rambus In-line Memory Module
RIP—Routing Information Protocol
RISC—Reduced Instruction Set Architecture
RJ45—Registered Jack 45
ROM—Read-Only Memory
RTX—Ray Tracing Texel eXtreme
RX—Radeon eXperience

SAN—Storage Area Network
SATA—Serial ATA
SBTV—Sistema Brasileiro de Televisão Digital
SCTP—Stream Control Transmission Protocol
SDK—Software Development Kit
SDLC—Software Development Life Cycle
SDRAM—Synchronous Dynamic RAM
SIM—Subscriber Identity Module
SIMM—Single In-line Memory Module
SIPP—Single In-line Pin Package
SMT—Surface-Mount Technology
SMT—Symmetric Multithreading
SoC—System-on-a-Chip
SO-DIMM—Small Outline DIMM
SO-RIMM—Small Outline RIMM
SPARC—Palo Alto Research Center
SQL—Structured Query Language
SRAM—Static Random-Access Memory
SSD—Solid-State Drive

SSL—Secure Sockets Layer
SSR SDRAM—Synchronous Dynamic Random-Access Memory
SUV—Sport Utility Vehicle

TB—Terabyte
TCP—Transmission Control Protocol
TDP—Thermal Design Power
TFT—Thin-Film-Transistor
TLS—Transport Layer Security
TN LCD—Twisted Nematic Liquid Crystal Display
TSMC—Taiwan Semiconductor Manufacturing Company, Limited
TV—Television

UDP—User Datagram Protocol
UEFI—Unified Extensible Firmware Interface
UHD—Ultra High Definition
UI—User Interface
UNIVAC—Universal Automatic Computer
UNIVAC LARC—UNIVAC Livermore Advanced Research Computer
USB—Universal Serial Bus
UV—Ultraviolet

VB .NET—Visual Basic .NET
VGA—Video Graphics Array
VLIW—very long instruction word
VMS—Virtual Memory System
VoIP—Voice Over IP
VPN—Virtual Private Network
VR—Virtual Reality
VRAM—Video RAM
VSI—VMS Software, Inc

WAN—Wide Area Network
Wi-Fi—Wireless Fidelity
WLAN—Wireless LAN
WWT—World Wide Telescope
WWW—World Wide Web

XDR—External Data Representation

References

WIKIPEDIA

https://en.wikipedia.org/wiki/Computer.
https://en.wikipedia.org/wiki/Computer_hardware.
https://en.wikipedia.org/wiki/Software.
https://en.wikipedia.org/wiki/Central_processing_unit.
https://en.wikipedia.org/wiki/Graphics_processing_unit.
https://en.wikipedia.org/wiki/Computer_memory.
https://en.wikipedia.org/wiki/Input/output.
https://en.wikipedia.org/wiki/Operating_system.
https://en.wikipedia.org/wiki/Application_software.
https://en.wikipedia.org/wiki/Computer_network.
https://en.wikipedia.org/wiki/Internet.
https://en.wikipedia.org/wiki/List_of_Intel_chipsets.
https://en.wikipedia.org/wiki/List_of_Intel_processors.
https://en.wikipedia.org/wiki/List_of_AMD_processors.

INTEL

https://ark.intel.com/content/www/us/en/ark/products/134599/intel-core-i912900k-processor-30m-cache-up-to-5-20-ghz.html.
https://www.intel.in/content/www/in/en/products/sku/90737/intel-celeron-processor-g3920-2m-cache-2-90-ghz/specifications.html.

AMD

https://www.amd.com/en/graphics/radeon-rx-graphics.

NVIDIA

https://www.nvidia.com/en-in/geforce/graphics-cards/30-series/rtx-3060-3060ti/.

GIGABYTE

https://www.gigabyte.com/Motherboard/GA-B250-HD3P-rev-10.
https://www.gigabyte.com/Motherboard/GA-B250-HD3P-rev-10#kf.

ASROCK

https://www.asrock.com/MB/AMD/WRX80%20Creator/.

RUFUS

https://rufus.ie/en/.

UBUNTU

https://ubuntu.com/.

STATISTA

https://www.statista.com/chart/24686/desktop-pc-market-share-in-the-us/.

NCES

https://nces.ed.gov/nationsreportcard/writing/lessons/performance.aspx.

WIKIMEDIA COMMONS

https://commons.wikimedia.org/wiki/File:Arpanet_1974.svg.
https://commons.wikimedia.org/wiki/File:Xerox_Alto_mit_Rechner.JPG.

LINUS TECH TIPS (YOUTUBE CHANNEL)

https://www.youtube.com/linustechtips.

SEIKO

https://museum.seiko.co.jp/en/collections/watch_latestage/collect021.

Index

Page references for figures are italicized.

Accelerated Processing Unit (APU), 25, 36
ACD/ChemSketch, 135
Acorn Computers, 9
active-matrix organic light-emitting diode (AMOLED), 18, 56
ADINA, 70
Adobe Acrobat Reader, 129, 134
Adobe Illustrator, 136
Adobe Photoshop, 71, 98, 136
Adobe Premiere, 70, 98, 136
Advanced Micro Devices (AMD), 15–16, 23, 35, 96
Advanced Research Projects Agency Network (ARPANET), 76, 77, 89–90
Advanced Technology eXtended (ATX), 15, 95, 117
AIM. *See* AOL Instant Messenger
AIO cooling system, 115, 118–19
AI. *See* Artificial intelligence
Algodoo, 71
algorithms, 66–68
Alphanumeric Television Interface Controller, 31
Altair 8800, 7
ALU. *See* Arithmetic Logic Unit
AM4 socket, 114

Amazon, 101, 103–4, 134
Amazon Echo, 19
AMD Crossfire, 119
AMD FX processors, 26
AMD Threadripper, 23, 27–28, 95–97, 115
AmigaOS, 61
AMOLED. *See* active-matrix organic light-emitting diode
analog computers, 8
Analytical engine, 8
Android OS, 61
antivirus apps, 91, 129
AOL Instant Messenger (AIM), 88
API. *See* Application Programming Interface
Apple, 21, 59, 61, 64, 100, 102, 105
Apple II, 60
AppleDOS, 60
Apple File System (APFS), 64
Apple Safari, 69
Application Programming Interface (API), 32, 35, 67
application software, 20, 65
apps, 68–71, *70*, 98, 127–29
APU. *See* Accelerated Processing Unit

147

Arithmetic Logic Unit (ALU), 5, 8, 13–14, 24
Arm64, 30
Artem Izmaylov Media Player (AIMP), 70, 128
Artificial intelligence (AI), 17, 19
AS/400, 60
ASRock, 103
Asus, 103
AT&T, 60, 77, 83, 90
Atanasoff, John Vincent, 9
Atari, 31, 61
Audacity, 71, 136
audio players, 69–70
audio ports, 96, 117
Autodesk Fusion 360, 71, 98, 136
Avast, 91, 128
AV-TEST, 91

Babbage, Charles, 8
Ball Grid Array (BGA), 28–29
Baran, Paul, 75
Basic Input/Output System (BIOS), 16, 21, 37, 60, 64, 120
Beginners' All-purpose Symbolic Instruction Code (BASIC), 2, 66–67, 133
Bell Labs, 21, 60
Berkeley Software Distribution (BSD), 59
Berry, Clifford E., 9
BestBuy, 101, 104
BGA. *See* Ball Grid Array
binary data, 1–2, 5
BIOS. *See* Basic Input/Output System
bit-length, 26
Blogger, 134, 137
blogging, 134, 137
Bluetooth, 17, 50
bootable USB drives, 122–23, *123–24*
booting, 21–22
Brain Computer Interface, 54
bridges, 78–79
BSD. *See* Berkeley Software Distribution

building computers, 107, *109–11*; case preparation, 113–14; motherboard preparation and installation, 114–18; powering on and BIOS setup, 120; safety precautions for, 112–13; tools for, 108–12
bus, 13, 15, 40
buying offline, 102–3
buying online, 103–5

C#, 67
cache memory, 24–25, 27, 32–33
CAD. *See* computer aided design
cascading style sheets (CSS), 85
cathode ray tubes (CRTs), 4, 11, 18
C/C++, 2, 19–20, 59, 65–66
CD/DVD drives, 19
central processing unit (CPU), 5, 7, 10, 13–14, 20, 23; architectures, 30; components, *24*, 24–26; generations of, 29–30; installing to motherboard, 114–16; operating system requirements and, 63–64; specifications of, 26–30; types of, 16
Cerf, Vint, 90
Cisco Internetwork OS (IOS), 61
CISC. *See* Complex Instruction Set Computer
clock speed, 26, 34
cloud computing services, 134, 137
cloud drives, 47
cloud storage, 134, 137
CLR. *See* Common Language Runtime
CMOS. *See* Complementary Metal Oxide Semiconductor
Colossus computer, 9
Commodore64, 60
Common Language Infrastructure (CLI), 20, 60
Common Language Runtime (CLR), 67
communication apps, 128–29
Complementary Metal Oxide Semiconductor (CMOS), 16, 37
Complex Instruction Set Computer (CISC), 30

computer aided design (CAD), 71, 95, 97, 136
computer networks, 73–81, *80*
computers, 1–2; analog, 8; architecture of, 5, *6*; building, 107–20, *109–11*; hardware components, 3–5, 13–19; high-end configuration, 99–100; history of, 8–9; instructions for, 2–3; low-end configuration, 99; offline buying, 102–3; online buying, 103–5; system checklist, 95–100; types of, 6–8
control flow, 8
Control Program/Monitor (CP/M), 60
Control Unit (CU), 5, 13–14, 24–25
cookies, 86–87
cooling fans, 17, 113, 115
CorelDraw, 71, 136
cores, 27, 32, 34
CP/M. *See* Control Program/Monitor
CPU. *See* central processing unit
CPU coolers, 119
CRTs. *See* cathode ray tubes
CSS. *See* cascading style sheets
CU. *See* Control Unit
CUDA/GPU cores, 32, 34

Data-Flow Diagrams, 67
Datagram Congestion Control Protocol (DCCP), 75, 78
data over cable service interface specification (DOCSIS), 85
DCCP. *See* Datagram Congestion Control Protocol
DDR SDRAM. *See* Double Data Rate Synchronous Dynamic Random-Access Memory
DEC. *See* Digital Equipment Corporation
Defense Communications Agency, 77
Dell, 100, 102, 105
Dennard, Robert H., 39
desktop computers, 6–7, 9–11, *10*, 120
DHCP. *See* Dynamic Host Control Protocol
Difference engine, 8

digital cameras, 3
Digital Equipment Corporation (DEC), 7, 60
digital subscriber line (DSL), 85, 89
Direct 3D, 32
Direct Memory Interface (DMI), 15
displays, 3–4, 11, 54–58, 98
DMI. *See* Direct Memory Interface
DNS. *See* Domain Name System
DOCSIS. *See* data over cable service interface specification
document scanner, 17
Domain Name System (DNS), 86
Double Data Rate Synchronous Dynamic Random-Access Memory (DDR SDRAM), 27, 32, 37, 41
DRAM. *See* dynamic random access memory (DRAM)
drivers, installing, 126
DSL. *See* digital subscriber line
Dynamic Host Control Protocol (DHCP), 85
dynamic random access memory (DRAM), 25, 38–39

EaseUS Todo, 128
eBay, 101, 103–4
educational apps, 71
EEPROM. *See* Electrically Erasable Read-Only Memory
86 DOS, 60
Electrically Erasable Read-Only Memory (EEPROM), 38
electromechanical analog computers, 8
Electronic Numerical Integrator and Computer (ENIAC), 9
electrostatic discharge (ESD), 17, 107
email, 87, 91
embedded dynamic random access memory (eDRAM), 34
Englebart, Douglas, 51
ENIAC. *See* Electronic Numerical Integrator and Computer
EPIC. *See* Explicitly Parallel Instruction Computing
ESD. *See* electrostatic discharge

Ethernet, 75, 80–81
Ethernet headers, 74
Ethernet network cards, 73
EXEC I OS, 60
Explicitly Parallel Instruction Computing (EPIC), 30

Faggin, Federico, 38
Fairchild Semiconductor, 9, 38
FAT32, 47, 64, 122, 127
FHD. *See* Full High Definition
fiber-optic internet connections, 85, 88–89
file systems, 47
File Transfer Protocol (FTP), 75, 77
firewalls, 79, 91
firmware, 64
5G wireless, 85, 88–89
flash memory, 38, 45
Floating-Point Unit (FPU), 24–25
Florida Virtual School, 131
flow charts, 67
FL Studio, 136
form factors, 15
FormScanner, 136
Forza Horizon, 93
FPU. *See* Floating-Point Unit
FreeBSD, 60, 63
Front-Side Bus (FSB), 15
FruityLoops Studio, 71
FSB. *See* Front-Side Bus
FTP. *See* File Transfer Protocol
Full High Definition (FHD), 18, 35

games, 93
Gates, Bill, 60
GDDR SGRAM. *See* graphics DDR synchronous graphics RAM
GeForce graphics cards, 31–33
GeoEnZo, 133
GeoGebra, 133
GEOS. *See* Graphic Environment Operating System
Gmail, 87
GOCR, 136
Google, 61, 86, 122, 132

Google Chrome, 69, 85, 129, 134, 137
Google Drive, 47, 134, 137
Google Earth, 133
Google Fiber, 83, 131
Google Nest, 19
Google Workspace, 137
GPT, 64
GPU. *See* Graphics Processing Unit
graphical user interface (GUI), 14, 17, 20, 31
Graphic Environment Operating System (GEOS), 60
graphics cards, 11, 119–20
graphics DDR synchronous graphics RAM (GDDR SGRAM), 39
graphics memory, 32–34, 39
Graphics Processing Unit (GPU), 5, 10, 13–14, 25, 31; components, 32–33, *33*; installing, 119–20; operating system requirements and, 64; specifications, 34–36
GUI. *See* graphical user interface

hacking, 90
hard disk drive (HDD), 14, 19, 97–98; cleanup and defragmentation, 127; components of, 42–44, *43*; specifications of, 44–45
hardware, 13–19
Harwell CADET computer, 9
HDMI. *See* High-Definition Multimedia Interface
HelenOS, 63
HeySpace, 137
High-Definition Multimedia Interface (HDMI), 18, 96
homework, apps and internet for, 132–35
Horizon, 63
HotSpot, 89
HP, 102, 105
HPE Cray, 8
HT. *See* hyper-threading
HTML. *See* Hypertext Markup Language
HTTP. *See* Hypertext Transfer Protocol

hubs, 78
Hypertext Markup Language (HTML), 85
Hypertext Transfer Protocol (HTTP), 75, 77
hyper-threading (HT), 28
HyperTransport/Uplink, 15

IA-64, 30
IaaS. *See* Infrastructure-as-a-Service
IBM. *See* International Business Machines
IBM 305 RAMAC, 42
ICQ, 69
ICs. *See* Integrated Circuits
IDE. *See* Integrated Development Environment
IGP. *See* integrated graphics processor
iMac, 100, 102
image files, 122
IMAP. *See* Internet Message Access Protocol
Infrastructure-as-a-Service (IaaS), 84
Inkscape, 133
in-plane switching (IPS), 18, 56
input devices, types of, 49–54
input/output devices (I/O devices), 3, 5, *10*, 15, 17, 49
installation media, 122–25
instant messaging services, 87–88
instruction sets, 30
Integrated Circuits (ICs), 9, 23
Integrated Development Environment (IDE), 65, 133
integrated GPUs, 35–36
integrated graphics processor (IGP), 25
Intel, 15–16, 23, 36, 96, 103
Intel 4004 processor, 23, 26
Intel 486DX2, 27
Intel Celeron, 27–28, 93, 97
Intel Core i3, 97
Intel Core i5, 97
Intel Core i7, 25, 93, 97
Intel Core i9, 23, 26, 28, 97
Intel Itanium, 30
Intel Pentium III, 28

International Business Machines (IBM), 7, 38–39, 42, 45, 60
Internet, 83
internet access, *88*, 88–90
Internet Explorer, 129
Internet Message Access Protocol (IMAP), 87
internet security, 90–91
Internet Service Provider (ISP), 83–84, *84*, 85–86, 89–90
I/O devices. *See* input/output devices
I/O interface, 5, 15, 19
IOS. *See* Cisco Internetwork OS
IP address, 73–74, 85, 90
IP headers, 74
IPS. *See* in-plane switching
Iris Xe, 36
ISO files, 121
ISP. *See* Internet Service Provider
iTunes, 70

J#, 67
Jamboard, 137
Java, 65–66, 85
joystick, 17

Kahn, Bob, 90
KERNAL, 60
keyboard, 3, 11, 17, 49–51, 98
Kilburn, Tom, 9, 38, 68
Kilby, Jack, 9

LAN. *See* Local Area Network
Land Grid Array (LGA), 28–29, 114
laptops, 7
LCD displays, 4, 11, 18, 54–58, 98
Learning Essentials for Microsoft Office On-Demand, 135
LED displays, 18
LenMus Phonascus, 133
LGA. *See* Land Grid Array
LibreOffice, 129, 133, 135
Linux, 21, 59, 61, 122, 125–26, 129
lithography, 29
Local Area Network (LAN), 75, 83
Logitech, 51

Index

Long Term Evolution (LTE), 85, 89
Lumen, 90

M.2 slots, 19, 45, 47, 96
MAC address, 73–74
MacOS, 20–21, 59–61, 63–64, 68, 122
Magix Music Maker, 136
Magix Sound Forge, 71, 136
Magix Vegas Pro, 71, 136
mainframe computers, 7, 9
malware, 90, 128–29
MAN. *See* Metropolitan Area Network
Manchester Baby computer, 9, 68
Man-In-The-Middle attacks (MITM attacks), 91
Masuoka, Fujio, 45
maximum resolution, 34–35
MCI Mail, 87
media development apps, 70–71
media player classic, 128
Media Player Classic—Home Cinema, 69
memory, 4–5, 8, 14, 37; bus, 40; cache, 24–25, 27, 31–33; cards, 19; controller, 39–40; flash, 38, 45; graphics, 32–34, 39; installing, 116; motherboard slots for, 96; operating system requirements and, 64; supported types of, 27; on system checklist, 97
Messenger, 69
Metal-Oxide-Semiconductor RAM (MOS RAM), 38
Metropolitan Area Network (MAN), 75
Micro-ATX, 15, 95
Microcenter, 101
microcomputers, 7, 9
microphone, 3, 17
Microsoft, 60
Microsoft app store, 127
Microsoft Excel, 98
Microsoft Math Solver, 71
Microsoft Office, 69, 98, 129, 133, 135
Microsoft Outlook, 128
Microsoft Photos, 128
Microsoft PowerPoint, 129

Microsoft Visual Studio, 71
Microsoft Windows, 20–21, 59, 61, 63, 68; image files for, 122; installing, 125–27
Microsoft Word, 98
MIDI. *See* Musical Instrument Digital Interface
Minecraft, 69
Mini-ATX, 15, 95
minicomputers, 7, 9
Mini-ITX, 15
Minix, 62
MITM attacks. *See* Man-In-The-Middle attacks
Mock, Owen, 60
modems, 89
monitors, 3–4, 18. *See also* displays
Moore, Gordon, 23
Moore's law, 23
MOS capacitors, 39
Moscow Center of SPARC Technologies, 30
MOSIX, 62
MOS RAM. *See* Metal-Oxide-Semiconductor RAM
MOS transistors, 39
motherboard, 10–11, 15, *16*, 17; choosing, 95–97; components of, *108*; preparing and installing, 114–18; size of, 95–96; socket types, 28–29
mouse, 3, 11, 17, 51–54, 98
Mozilla Firefox, 68, 85, 129, 137
Mozilla Thunderbird, 87, 128
MS-DOS, 60
multimedia apps, 128
MURAL, 136
MuseScore, 133
Musical Instrument Digital Interface (MIDI), 69–70, 136
MySQL, 67, 69

NAS. *See* Network-Attached Storage
National Foundation Network (NSFNET), 90
Near-Field Communication (NFC), 75

Nero Burning ROM, 121–22
.Net, 65, 67
NetBalancer, 128
Network-Attached Storage (NAS), 47
networking devices, 78–79
Network Interfaces Controller (NIC), 73
network protocols, 75
network topology, 79–81, *80*
New Technology File System (NTFS), 47, 64, 127
NeXTSTEP, 60
NFC. *See* Near-Field Communication
NIC. *See* Network Interfaces Controller
Nintendo, 3, 7
non-volatile memory, 37–38
non-volatile memory express (NVMe), 45, 47
non-volatile storage devices, 19
Norman, Robert H., 38
Noyce, Robert, 9
NSFNET. *See* National Foundation Network
Nvidia, 31–32, 35, 97, 119
NVMe. *See* non-volatile memory express

Objective-C, 65
Object-Oriented Programming (OOP), 66–67
OCR. *See* Optical Character Reader
Office apps, 129
OMLED. *See* organic light-emitting diode
OMR. *See* Optical Mark Reader
One Time Programmable (OTP), 38
online storage, 47
OOP. *See* Object-Oriented Programming
OpenCL, 32
OpenGL, 32
Open Source Physics, 135
Open Textbook Library, 134
OpenVMS, 60
Operating System (OS), 13, 20, *21*, 59–62, 95; architecture of, *62*, 62–63; installing, 121–27; system requirements, 63–64

Optical Character Reader (OCR), 136
optical discs, 14–15, 121, 124–25
Optical Mark Reader (OMR), 136
Oracle, 67
organic light-emitting diode (OMLED), 18
OS. *See* Operating System
OS/2, 61
OS/360, 60
OS/400, 60
OSI model, 77–78
OTP. *See* One Time Programmable
output devices, 3–4, 54–58
overclocking, 26

PaaS. *See* Platforms-as-a-Service
Palo Alto Research Center (PARC), 51, 60
Parallel ATA (PATA), 19, 44
parallel ports, 18
PARC. *See* Palo Alto Research Center
Passive Optical Network, 85
PATA. *See* Parallel ATA
Paterson, Tim, 60
Patrick, Robert L., 60
PCIe. *See* Peripheral Component Interconnect Express
PDF viewers, 129, 134
PDP-7, 60
PDP-8, 7
PDP-11, 60
PerfectDisk, 127
Peripheral Component Interconnect Express (PCIe), 35, 45, 47, 96, 119
PGA. *See* Pin Grid Array
phishing, 91
photo editing, 136
Pineapple Pictures, 128
Pin Grid Array (PGA), 28
Platforms-as-a-Service (PaaS), 84
PlayStation, 3, 7, 31
Point of Presence (PoP), 83
POP. *See* Post Office Protocol
POST. *See* Power-On Self-Test
Post Office Protocol (POP), 87
power consumption, 29, 34

Power-On Self-Test (POST), 21
Power Supply Unit (PSU), 113, 117–18
printers, 4, 18
Programmable Read-Only Memory (PROM), 38
programming languages, 20, 65, 133
Project Chrono, 135
PROM. *See* Programmable Read-Only Memory
PSU. *See* Power Supply Unit
Python, 2, 19, 65–67, 85

QBASIC, 133
QuickPath Interconnect (QPI), 15

Radeon graphics cards, 35, 97
Radio Corporation of America (RCA), 55
Radio Shack Color Computer, 61
Random Access Memory (RAM), 5, 14, 38, *39*, 39–42, 97, 116
ransomware, 90
Raspberry Pi, 120
RCA. *See* Radio Corporation of America
RDMS. *See* Relational Database Systems
Read Only Memory (ROM), 5, 14
Reduced Instruction Set Architecture (RISC), 30
RegCleaner, 128
register, 24–25
Relational Database Systems (RDMS), 67
resolution, 34–35
Revo Uninstaller, 128
RISC. *See* Reduced Instruction Set Architecture
ROM. *See* Read Only Memory
routers, 79
Rufus, 121–22

SaaS. *See* Software-as-a-Service
SAGE. *See* Semi-Automatic Ground Environment

Samsung Electronics, 39
SAN. *See* Storage Area Network
SanDisk Corporation, 45
SATA. *See* Serial ATA
satellite internet access, 89
Scalable Link Interface (SLI), 119
scanner, 17
Schadt, Martin, 55
Schmidt, John, 38
SDK. *See* Software Development Kit
SDLC. *See* Software Development Life Cycle
Seagate, 42
security apps, 128
Seiko, 55
Semi-Automatic Ground Environment (SAGE), 77
Serial ATA (SATA), 19, 44–45, 96, 117
7zip, 128
SIM. *See* Subscriber Identity Module
simple mail transfer protocol (SMTP), 87
simulation apps, 70
Skype, 69, 87, 128, 136
SLI. *See* Scalable Link Interface
SMT. *See* Surface-Mount Technology; symmetric multithreading
SMTP. *See* simple mail transfer protocol
socket types, 28–29
SoCs. *See* system-on-a-chip ICs
software, 19–22, 98
Software-as-a-Service (SaaS), 84
software development apps, 71
Software Development Kit (SDK), 65
Software Development Life Cycle (SDLC), 68–69, *70*
solid state drive (SSD), 19, 45–46, *46*, 47, 97–98
Sony, 60
sound cards, 11
Soundfont Midi Player, 70
spamming, 91
speaker systems, 4, 11, 18–19
Sprint, 90
spyware, 90

SQL. *See* Structured Query Language
SRAM. *See* static RAM
SSD. *See* solid state drive
SSE. *See* streaming SIMD Extensions
SSR SDRAM. *See* Synchronous Dynamic Random-Access Memory
Starlink, 89
static RAM (SRAM), 38
Steinberg Nuendo, 71
Stellarium, 135
Storage Area Network (SAN), 75
storage devices, 4–5, 37, 64, 97–98, 113–14, 127. *See also* hard disk drive; solid state drive
StorageTek STC 4305, 45
streaming SIMD Extensions (SSE), 26
Structured Query Language (SQL), 67
Subscriber Identity Module (SIM), 85, 89
supercomputers, 7
Surface-Mount Technology (SMT), 29
symmetric multithreading (SMT), 28
Synchronous Dynamic Random-Access Memory (SSR SDRAM), 32, 39
system checklist, 95–100
system-on-a-chip ICs (SoCs), 9
system software, 20

tablets, 7
Taiwan Semiconductor Manufacturing Company Limited (TSMC), 31
Taylor, Robert, 77
TCP headers, 74
TCP/IP, 75, 77–78, 83, 90
TDP. *See* Thermal Design Power
teaching, apps and internet for, 135–37
Telegram, 69
telephone lines, 88–89
Teller, Edward, 7
Tetris, 69
Texas Instruments, 9
Thermal Design Power (TDP), 29
Thomson, William, 8
TimeCamp, 137
tools, for building computers, 108–10

Tootill, Geoff, 9
Tor, 90
Torpedo Data Computer, 8
Torvalds, Linus, 21
Toshiba, 38, 45
TR4 socket, 115
transistors, 9, 23
Transmeta, 30
Trojan horses, 90
TSMC. *See* Taiwan Semiconductor Manufacturing Company Limited
Tutors.com, 137

Ubuntu, 21, 125–26, 129
UDP. *See* User Datagram Protocol
UEFI. *See* Unified Extensible Firmware Interface
uGet, 129
UI. *See* user interface
Ultra High Definition (UHD), 36
Unified Extensible Firmware Interface (UEFI), 64
uniform resource locator (URL), 85
Universal Automatic Computer (UNIVAC), 7, 60
Universal Serial Bus (USB), 17–18, 53, 96, 117
Unix, 21, 59
URL. *See* uniform resource locator
USB. *See* Universal Serial Bus
USB flash drives, 121–23, *123–24*, 124–25
User Datagram Protocol (UDP), 75, 78
user interface (UI), 71
utility apps, 128

Van-Basco's Karaoke Player, 70
VAX, 60
vector editing apps, 133, 136
Verizon, 90
Very Long Instruction Word (VLIW), 30
VGA. *See* Video Graphics Array
video editing, 97, 136
video games, 69

Video Graphics Array (VGA), 18
VideoLAN Client (VLC), 69
videotelephony apps, 134
Virtual Chemistry Lab, 133
Virtual Private Network (VPN), 75, 89
virtual schools, 131
viruses, 90–91
Visual Basic.NET, 67
VLC. *See* VideoLAN Client
VLIW. *See* Very Long Instruction Word
Vlogging, 137
Voice over IP (VoIP), 69, 77, 128, 134
voice recognition, 17
VoIP. *See* Voice over IP
VPN. *See* Virtual Private Network
Vulkan, 32

Walmart, 101, 104
WAN. *See* Wide Area Network
web browsers, 69, 129, 137
webcam, 17
webpages, 86–87
website addresses, 85–86
Western Digital, 42, 47
WhatsApp, 69, 88, 91
whiteboard apps, 136–37
Wide Area Network (WAN), 75, 77, 89

Wi-Fi, 17–18, 50, 75, 91
Wikibooks, 134
Wikipedia, 134
Williams, Frederic C., 9, 38
Windows 11, 28, 64, 98
Windows 98, 28
Windows CE, 62
Windows Image Viewer, 128
WinZip, 128
wireless internet connections, 85
wireless LAN (WLAN), 81
workstations, 9–11
World Wide Telescope, 135
World Wide Web (WWW), 85–86

x86-64, 30
Xerox Alto, 51, 60

Yahoo messenger, 88
Yamaha S-YXG50, 70
YouTube, 135, 137

Z2 computer, 8
zero insertion force socket (ZIF socket), 28
Zircon, 63
Zoom, 134, 136
Zuse, Konrad, 8

About the Author

Debojit Acharjee is a computer professional and also a computer enthusiast. He was passionate about computers since his childhood, and he learned about computer hardware like digital logic gates, microprocessors memory and so on, when he was only a school-going kid. He also acquired good knowledge about software since he used a computer during his school days, and was able to create a chat bot using BASIC programming language.

Besides having qualifications in computer software, he also possesses a government certificate (with A+ grading) in computer hardware and networking, and has good skills in building a desktop computer.

Debojit is also a writer and has written many articles on computer technologies like Artificial Intelligence (AI), blockchain, DevOps, Python, and so on. He believes that a perfect combination of hardware and software can make a computer worthy.

www.ingramcontent.com/pod-product-compliance
Lightning Source LLC
Chambersburg PA
CBHW022014300426
44117CB00005B/186